moveable feasts

Sarah Murray is a *Financial
Times* contributor. She has lived
and worked in Hong Kong,
Vietnam and South Africa
and now divides her time
between London and New York.
This is her first book.

TO MY PARENTS

moveable feasts

THE *INCREDIBLE JOURNEYS* OF THE *THINGS WE EAT*

SARAH MURRAY

First published in Great Britain
2007 by Aurum Press Limited
7 Greenland Street
London, NW1 0ND

A catalogue record for this book is available from the British Library.

ISBN-10: 1 84513 234 3
ISBN-13: 978 1 84513 234 7

1 3 5 7 9 10 8 6 4 2
2007 2009 2011 2010 2008

Design: Roger Hammond

Typeset by SX Composing DTP, Rayleigh Essex
Printed and bound in Great Britain by Creative Print and Design,
Ebbw Vale

contents

Introduction   1

1   Liquid Gold: The ancient amphora delivers Roman riches   7
2   The Business of Bones: The Norwegian salmon pays
      a visit to China   33
3   Cannon Fodder: Battlefield food fuels packaging
      technology   60
4   Plane Fare: The Berlin Airlift secures a Cold War victory   82
5   Tiffin Travels: Curry catches the corporate imagination 107
6   Yes, We Can Do Bananas: Refrigerated ships shape Central
      American regimes   134
7   Whey to Go: Mongolian nomads practise mobile
      biochemistry   161
8   Barrels and Bouquets: The oak tree leaves its mark   191
9   A Quick Cuppa: Commercial competition speeds the racing
      tea clippers   211
10   Food with Altitude: Jet planes dispatch a strawberry
      for all seasons   238
11   Elevated Design: Buffalo grain feeds the Bauhaus
      inspiration   262
12   Going with Gravity: Cold War weaponry finds a
      new purpose   285

Epilogue   307
Acknowledgements   316
Bibliography   317
A Note on Sources   325
Index   326

introduction

PLUCK A CLUSTER of green peppercorns from a vine, put one of the tiny berries in your mouth, crunch it between your teeth, and something curious happens. A small explosion of flavour travels at speed through your body, first around your head and then through veins and arteries (or so it seems). It is a remarkable sensation. Pain and pleasure intermingle in a heady rush of piquancy. It is not only pepper's flavour that has a propensity to travel, however. As a valuable commodity, this king of spices has been shipped around the world for centuries, covering tens of thousands of miles in the hands of everyone from Arab traders and Venetian merchants to European colonisers. Vasco da Gama, the Portuguese explorer, battled for two years across 27,000 miles of ocean in his search for 'Christians and spices', landing on India's tropical Malabar Coast in 1498. There, he found the spices, if not the Christians, and soon after, an ancient global trade was brought under

Portuguese control, fuelling the growth of the first European empire in Asia.

Today, pepper is conveyed in industrial-size shipping containers on cargo vessels, trains and trucks. As an ingredient in so many other foods – which in turn cover enormous distances – it remains one of the best travelled of edible commodities. However, the voyages made by pepper differ little from those embarked on every day by much of what we eat. Grapes bound for Des Moines, Iowa, might have been on a long and complicated voyage from Chile by the time they end up on the supermarket shelf, heading north by ship to the port at Los Angeles or through the Panama Canal and up the east coast of America to Philadelphia, where after a short layover they are trucked several thousand miles across several US states. After a long and arduous climb up a wooden trellis, French beans may well feel that they have travelled quite far enough. That is unless they were grown in Kenya, in which case they will be rudely plucked from their vine, blast-chilled, packed inside a refrigerated container and flown thousands of miles by jet plane to London. In cities such as Beijing and Shanghai, affluent consumers now enjoy everything from Italian olive oil to Japanese noodles, Belgian chocolates and French cheeses, all of which have undergone lengthy journeys before reaching their destination. In fact, most of what we consume travels thousands of miles from its origins to its grand finale on the dinner table. How on earth did this happen?

Actually, globetrotting fodder is nothing new. Long before Vasco da Gama's arrival in India, food was being hauled over great distances. The ancient Silk Road, the trading conduit for, among other goods, tea and spices, was at least 7,000 miles long. The Romans moved food all over their empire – British archaeologists recently discovered a 2,000-year-old food jar with a label showing it once contained fish paste thought to have been shipped to Carlisle from Cadiz, Spain, more than 1,400 miles away. Some might ask how Roman officers in a Celtic corner of ancient Britannia acquired a taste for fish paste, or what accompanied it at dinner. My question would be how exactly did the paste get to such a remote corner of the Roman world? What sorts of containers did it travel in and how, in the absence of refrigeration and vacuum-sealed packs, did the fishy concoction not go bad during its journey?

Food has always been among the more difficult commodities to move around. Unlike other tradeable goods, it is often liquid, messy or perishable. Shipping it across the world has challenged the ingenuity and technical resources of engineers and inventors from the earliest times. As a result, the things we eat have travelled inside everything from humble wooden barrels to atmosphere-controlled shipping containers; from sleek nineteenth-century tea clippers with sails billowing at the mast to wide-bodied jet planes travelling at 500 miles an hour. Collectively, these vessels and machines underpin the global food system. They are the essential tools in getting things from farm to fork.

Some trips appear at first glance to make no sense at all. Today, fish is often frozen and sent on a ship to China where it is defrosted for filleting before being refrozen and returned to the USA and Europe. Yet this improbable round trip makes perfect sense to fish sellers because container shipping and high-tech refrigeration have combined to make ocean-borne transport so cheap and efficient that they can cut their labour costs by processing fish on the other side of the world. Such bizarre journeys are not what we usually think about when

introduction

browsing the supermarket aisles. But these are exactly the sorts of voyages this book will describe.

Shifting food around has at times altered the nature of the food itself. Farmers in Japan have developed a means of 'training' watermelons to grow in square form so they are easier to pack, move, stack and store. In use for millennia, the barrel is a sturdy, watertight container that is easy to roll, despite its weight when full – but it has also been responsible for changing the taste of its contents. For winemakers, the barrel is now a powerful element in the creative palette. Alcoholic drinks such as wine, rum, whisky and port have all had their flavours manipulated by the oak staves of this efficient moveable container. Foods, too, have benefited from storage and transit in wood. Balsamic vinegar is, like wine, aged in oak barrels. Fish produced by the 400-year-old Cornish pilchard industry once relied for its flavour on being cured in dry salt then pressed to extract the excess oil and water and packed in pine casks before being shipped, mainly to Italy, where *Salacche Inglese della Cornovaglia* added a tangy bite to plates of polenta.

But the story of transporting food goes beyond the filling of stomachs. Take those peppercorns. Vasco da Gama might have found India woefully lacking in Christians but, while profiting from the spice trade, the Portuguese soon took up the task of converting the locals, spreading the religion into Asia. In Central America, seismic shifts in the political landscape were brought about when refrigerated steamships helped transform an unknown tropical curiosity into 'the poor man's fruit', giving US companies the economic clout with which to manipulate the 'banana republics' from which they harvested the yellow fruit of what is actually a giant herb. In 1842, when a nineteenth-century entrepreneur called Joseph Dart introduced a system of mechanical elevators that speeded up the transhipment of grain in Buffalo, New York, he helped transform the city's fortunes. What he did not know was that years later, the mammoth structures housing his remarkable machine would provide sparks of inspiration for an entirely new theory of architecture: modernism.

Many things have shaped the world – science, democracy, art, war, philosophy, politics – but the odysseys of food have rarely made it into the history books. *Moveable Feasts* will redress the balance. It will show that the movement of food, often over vast distances, has for centuries been part of human life – an inevitable consequence of the quest for sustenance. It will explore some of the complex technologies and systems used to deliver dinner. It will survey mankind's ingenuity (driven at every turn by his stomach) in devising ways of shifting sufficient quantities of the things we eat and drink from their origins to our tables. It will romp through history in search of eccentricities in the moveable food chain, whether in Italy, Spain, India, or Mongolia. It will demonstrate that the ability to send food across international borders benefits not only those sitting at the dinner table but also farmers such as the Kenyans who are now selling their beans to London supermarkets. It will also look at some of the complex tradeoffs that emerge as we try to ensure that our food supply – one that now relies heavily on fossil fuels – is sustainable.

First, let's go shopping. Basket in hand, I have picked up a series of items, one of which will appear in each chapter. Some are protagonists in the modern food industry. Others are part of centuries-old trade routes. Now, where was I? Oh yes, olive oil. Ranks of bottles confront me on the shelves – everything from a chic, minimalist brand from Greece to Tuscan bottles with elaborate old-world labels. But I am going for the Extra Virgin from Andalusia in southern Spain. Think about it for a moment – I am about to take home with me a liquid that has been crushed out of olives grown on a Mediterranean grove far away from this supermarket. By the end of my shopping trip, I will also have in my basket a pound of Norwegian salmon, a tin of plum tomatoes from San Marzano, Italy, a packet of chewing gum, a pre-prepared vegetarian curry, a bunch of Guatemalan bananas, a tub of low-fat yoghurt whose origin is hard to determine, a bottle of Californian chardonnay, a box of tea plucked from bushes in southern China, a carton of Spanish strawberries, a packet of flour milled in the USA and a corn on the cob.

Collectively, the items in my basket have travelled tens of thousands of miles before reaching the supermarket. More to the point, they have left a trail of footprints in their path, stirring up economic, social and political change along the way. For some, the phrase 'moveable feasts' refers to religious holidays, such as Easter or Chinese New Year, that fall on different calendar days each year, their dates governed by anything from theological calculations to the position of the moon. For others, a moveable feast is a task or decision that can be altered to suit changing circumstances. This book will use a more literal interpretation: the idea that the things we eat and drink are eminently (and sometimes against all odds) moveable.

1

Liquid Gold

The ancient amphora delivers Roman riches

Olive oil: Oil derived from the fruit of the olive tree

Origin: The Mediterranean region

Etymology: Latin *oliva*, from Greek *elaia*

Legends: According to the Romans, Hercules was given the task of spreading olive trees. Roaming the Mediterranean shores with his olive staff in his hand, he would strike the ground, sending out roots from which trees then grew

José Remesal Rodríguez holds a piece of pottery up to the sunlight. He is standing at the top of Monte Testaccio, a small, unassuming hill on the southern fringe of the Aventine, a short ride from Rome's city centre and within sight of some of Europe's greatest monuments. The Pyramid of Cestius and the Protestant Cemetery are nearby. Further in the distance, the majestic dome of the Pantheon, Borromini's extraordinary spiral tower at the Church of St Ivo and the pompous Monument to Victor Emmanuel II rise up above the low-slung buildings of the city. It is an impressive display – a visual excursion through Italian history from Roman times via the Renaissance and on to nineteenth-century unification. But the professor is not paying much attention to the view. He is too busy examining the chunk of clay in his hand. It is pale brown and bears a deep mark that appears to have been stamped into the clay while wet. There is nothing refined about this thick fragment of earthenware. Its form is clumsy; its surface rough. It was clearly not part of any sort of decorative or ceremonial object. It is in fact a piece of a Roman transport amphora – a ceramic pot about the size of a small barrel that almost 2,000 years ago carried food to Rome, capital of an empire stretching from the lowlands of Scotland to the deserts of Africa, from Spain to the Persian Gulf.

'It is Baetican, of course,' pronounces the professor. Baetica is today's Spanish province of Andalusia, the southernmost region of the Iberian Peninsula, home to flamenco and known for a simple, unpretentious cuisine that includes gazpacho, fried fish and cured ham. Spectacular Andalusian architecture such as the Mezquita in Córdoba, a cathedral that was once a mosque, provides a reminder of the presence of the Moors, the Muslims who ruled from the eighth century to the fifteenth. Long before that, however, the Romans were in charge. That was when Baetica was part of Hispania – an area now occupied by Spain, Portugal, Andorra and Gibraltar. Roman soldiers first arrived there in 218 BC and, as direct imperial rule was established, Hispania became a prized part of the empire. Three emperors, Trajan, Hadrian and Theodosius I, would be born there.

As one of Hispania's imperial provinces, Baetica was an important source of food for the empire. And this dry, mountainous swathe of land was the starting point for the piece of pottery Remesal is holding. The fragment is part of a second-century transport jar that set out to Italy from a vast agricultural estate owned by a wealthy senator at a time when Rome was at the height of its power. Produce from this fertile land would have been loaded into the jar, heaved on to a vessel by bonded labourers and shipped to Rome. There, it ended up in the homes and palaces of everyone from philosophers and politicians to manual workers and freed slaves. It may be small and dusty, but this fragment of pottery is part of the endless patchwork that is the history of the Roman Empire. The story it has to tell is one of immense wealth built on trade in an essential commodity: olive oil.

Gathered around the professor on Monte Testaccio are students and archaeologists who have travelled from the USA, Spain and Poland to join Remesal in his ambitious archaeological investigation of this modest-looking hill. It is not the best day for it. On an uncharacteristically soggy October morning, most of those present are wrapped up in brightly coloured plastic raincoats and fleece jackets. Umbrellas are at the ready. Below, the hum of traffic is accompanied by a cacophony of barking dogs, crowing roosters and pealing church bells as the city slowly wakes. A rainbow arches briefly across the tempestuous sky as the sun attempts to break through the clouds. Softly, the rain starts to fall.

A spot of bad weather does not trouble the professor, however. He is far too interested in what lies below his feet to worry about what is happening in the sky. A bearded, bespectacled Spaniard who seems at his happiest with a cigarette in one hand and a piece of pottery in the other, Remesal has spent the past couple of decades uncovering the stories hidden beneath Monte Testaccio's grassy slopes. This is his stomping ground and in blue jeans and khaki safari jacket, he looks entirely at home clambering over the uneven ground on the broken pieces of Roman amphorae scattered underfoot. 'I spend a month here each year and every time we come, we find something different,' he says, speaking in heavily accented French. 'I've got to know this hill pretty well, but there are always surprises.' Remesal talks with a deep, gravelly voice. It sounds as if, over the years, particles of dust from the pots he studies have become lodged in his throat.

The archaeological dig on Monte Testaccio is like no other. While most archaeologists spend days scrabbling about in the dirt in the hopes of finding something of interest, here they have no such worries. This is because the entire hill is made of archaeological material – there *is* no dirt here. What lies beneath the thin layer of topsoil is nothing but millions of broken pots. Each year, Remesal and his colleagues come here to carve a large square pit about ten feet deep into this archaic mound and study what they excavate. Like the rings of a tree, each layer of pots corresponds to a moment in time. In this year's pit, they have got down as far as the reign of Marcus Aurelius, around AD 175. As the dig progresses, Italian contract workers in white plastic hard hats stand at the bottom of the hole, chipping carefully away at it and filling buckets with pieces of amphora. Up on the surface, colleagues use a rope to heave load after load of fragments out into the fresh air and over to the centre of activity – a collection of large plastic tubs filled with muddy water around which students and academics sit and gossip as they wash 2,000-year-old layers of dust from the chunks of earthenware in their hands.

It is a busy scene. Dotted about the place are bright orange, yellow and green plastic crates into which the shards are thrown – one box for each category of fragment. Some boxes contain those with 'form' (handles, necks or bases). Others store pieces on which stamped marks, rough scratches or painted inscriptions are visible. Then there are boxes for the bulk of the pieces – shards without recognisable shape or markings known as 'no form'. At a large table, several archaeologists are working on what must be one of the world's more difficult jigsaw puzzles as they try – mostly in vain – to recreate entire pots by fitting together some of the larger pieces retrieved from the same excavation level.

At the end of the dig, much of what has been heaved up from below the hill's surface will be thrown back into the hole. It is a strange phenomenon. A single one of these shards found in a field anywhere else would generate great excitement among historians and archaeologists. Here on Monte Testaccio, however, pieces of Roman amphorae are being heaved up from below the hill's surface in bucket loads throughout the day. After washing the fragments, the volunteers casually throw them into the plastic boxes as if they were vegetables, scrubbed and ready for cooking. It looks as if they are preparing for a mammoth vegetarian feast.

Monte Testaccio is one of the world's more curious ancient relics. It is actually a vast rubbish heap. For more than two centuries, olive oil amphorae were dumped here after their contents had been unloaded and distributed to consumers. As Romans used up more and more oil, the pile of shards grew, creating over two centuries a hill made up entirely of bits of pottery. Bases, handles, rims, necks and body fragments all ended up here. Roman emperors came and went, battles were won and nations beaten into submission, but the mountain of pots kept rising. Some of the amphorae were shipped from North Africa, but at least eighty per cent of Monte Testaccio's unlikely treasures originated in Baetica. 'We are standing on Spanish territory,' declares Remesal with a grin. And he is right – what lies below our feet is a gargantuan mound made from the clay of southern Spain, a chunk of foreign soil that ended up on Italian shores.

Today, few notice this bizarre monument to ancient Rome's commercial might. After all, it is hardly located in the most glamorous setting. The modern district of Testaccio is one of the city's seedier areas, now famous for taverns, gay bars and nightclubs with names such as Caffè Latino Jazz Club, On The Rox and Chattanooga – places where Rome's night owls like to spend their time and money. But even these trendy establishments have their history rooted in the mountain of jars around which they cluster. In the Middle Ages, the hill became the focus for all kinds of religious festivals and secular revelries and a collection of taverns and restaurants opened around Monte Testaccio's slopes. From the seventeenth century, wine merchants who dug caves out of its slopes found that the dense ceramic make-up of the hill provided a cool atmosphere that was ideal for storing wine. Today, disco music thuds out from the caves at the base of the hill and, in packed restaurants, hungry diners enjoy the oxtail stew and the sweetbreads that have long been popular in this district, which was once home to the city's main slaughterhouse. Meanwhile, at the back of several of the restaurants, the pot fragments quietly watch over the proceedings, clearly visible behind glass walls.

Dining and dancing might be the order of the day now, but in the first and second centuries, all activity around here centred on waste disposal. There are several theories as to how it was organised. Some believe that empty pots were hauled up the ever-expanding hill by mules – each animal might have carried about four amphorae – and were broken up at the summit. Others speculate that the jars were smashed below before being taken up to their final resting place. From time to time, lime was poured on the broken shards to counter the smell of rancid oil and to prevent the spread of disease. With each cargo vessel that arrived on the banks of the Tiber, the hill grew bigger. Today it is, by some accounts, about 165 feet high and takes about twenty minutes to walk round. True, it is not much to look at compared with the Colosseum, the giant showcase for Roman cruelty, or the Pantheon with its mighty dome, but this man-made mound is one of the world's most important archaeological sites – a critical corner of ancient Rome, offering a remarkable window on to the empire's economic life.

The nineteenth-century writer Rodolfo Lanciani certainly appreciated the significance of old trash. 'The hill itself may be called a monument of the greatness and activity of the harbour of Rome,' he wrote in an 1897 description of Monte Testaccio, a quarter of a century after an Italian priest, Father Luigi Bruzza, and Heinrich Dressel, an Italian-Prussian professor, began to excavate the site. On a frosty morning in January 1872, the pair climbed up the slopes and set about analysing what they found there. By the time their work was complete, Dressel had scrutinised and set down details of more than 3,000 marks stamped on to handles of the amphorae, as well as nearly a thousand inscriptions written on the body of the pots by Roman insurance agents, ship's captains or customs officers. The amphorae assessed by Dressel's project represent a fraction of the hill's historical data. More than fifty million pots found their final resting place here. Today, while teams excavate more than 500 cubic feet of material each season, they are barely scraping the surface of this remarkable dump.

Not all Roman amphorae were disposed of as they were here, in a great big pile. Elsewhere, in early examples of recycling, they were used for storage in domestic kitchens and warehouses, and even as urns containing the ashes of the dead. Smashed into pieces, they became part of the fabric of buildings. Packed into the rubble core of walls, they acted as insulation. In roofs and domes, they helped lighten the structure's weight, and in the walls of theatres, the curved fragments enhanced the acoustics, amplifying the sound of music and voices. In short, these pots were extremely useful. So the fact that a mountain of amphorae accumulated in Rome, broken up, unwanted and collecting dust, indicates the scale of demand for olive oil in the first and second centuries.

Rome was then the largest city in the ancient world with a population of about a million. It was made up mainly of residences. It was tremendously crowded. It had no industrial manufacturing or food production of its own to speak of. Romans were consumers, not producers and most of what they ate had to be brought in from other parts of the empire – in extremely large quantities.

Olive oil was among the most important of the imports. An extraordinarily nutritious food product containing edible fats and high levels of vitamins A and E, olive oil was used by the Romans to fry, bake and roast their food. It was a key ingredient in bread as well as in salad dressings. By night, olive oil lamps provided lighting in domestic households, temples, baths and palaces. At sporting events, athletes smeared themselves with oil before competing, and olive oil was the base for most perfumes and cosmetics. Each Roman citizen probably consumed up to 13 gallons of the liquid a year (by comparison, Italians today use 4.5 gallons a year) and larger homes or taverns stored hundreds of gallons in a *dolium*, a huge jar that was dug into the ground to provide a cooler storage vessel. So Monte Testaccio's amphorae fragments are evidence of consumption on a massive scale. Some estimate that its well-travelled pots would collectively have carried an astounding 1.6 billion gallons to their destination (the same amount of liquid that would be generated by flushing a toilet once a second for thirty-two years).

Monte Testaccio has more secrets to reveal. A wide variety of markings on the pottery shards make the hill rather like a giant accounts book detailing the export and import of olive oil. Instead of ledgers recording income, expenditure and accounts receivable, stamps, scratches and painted inscriptions tell us of the estates producing the oil, the companies that shipped it and the customs officials in Spain and Rome who checked the goods on departure and arrival. The painted inscriptions are the most intriguing of the marks. The beauty of these strange and ephemeral hieroglyphs, with their elaborate flourishes and curls, is made more entrancing by the rarity of their occurrence elsewhere. When such pottery is found at other sites, exposure to light or moisture means any ink inscriptions have long since disappeared. At Monte Testaccio – a giant time capsule in which every shard has been protected and kept dry by a layer of topsoil and grass – details remain that allow us to trace each stage of a pot's passage from the olive estate to the docks of the Tiber. The month a particular pot left Spain can be pinned down. The exact date it arrived in Italy is also recorded.

Here in Rome, we learn of the fortunes of the Spanish businessmen who profited from this valuable liquid. When a group of shards bearing the same markings are found at the same level, it becomes possible to start building a picture of certain families and the years in which their olive estates produced good harvests. The Baetican landowners and merchants made wealthy by olive oil had names like MM. Iulii of Astigi, a family at the heart of the trade for more than three generations, or M. Iulius Hermes Frontinianus, whose son M. Iulius Hermesianus followed his father into the business. It is often unclear whether these people were traders or producers, but what is certain is that they profited from olive oil in some way. Shipped across the empire, olive oil turned Baetica into Hispania's wealthiest province, with architectural, political and social structures that emulated those of Rome itself. Common coinage was introduced. Bridges and aqueducts were constructed. Latin became the province's official language.

The landowners and businessmen immortalised by Monte Testaccio's ceramic fragments lived in fine villas and led lives of luxury, surrounded by dozens of slaves and attendants. They hosted expensive private parties, staged theatrical shows, purchased works of art and erected funerary monuments to themselves. Some Baetican oil traders made it into the highest echelons of the aristocratic elite, particularly if they used their spare cash to make donations to the *collegia*, voluntary associations that were part of the social fabric of the Roman Empire. A Narbonne-based Spanish olive oil dealer, Sextus Fadius Secundus Musa, was one such figure. A prominent dealer whose amphorae have been found on Monte Testaccio, Sextus would have possessed numerous properties near his olive estates, all decorated with elaborate mosaics and sculptures and used as venues for lavish social events. He was clearly a well-known philanthropist, giving generously to the *collegia*, for the city of Rome's council rewarded his patronage with the erection of a statue. Money did not always lead to social prominence in the Roman world but it certainly helped.

None of this would have been possible without the assistance of an unassuming clay jar. The shape of this simple but cleverly designed container – a cross between an egg and a torpedo – made it remarkably strong and easy to pack into a vessel's hold. The curve of the pot's side fitted snugly against the curve of the ship. Its pointed base allowed the jar to fit neatly in between the shoulders of the amphorae in the row below, preventing the cargo from rolling around during transit. The base also served as a third grip – supplementing its two handles – for dockworkers to grasp during the unloading process and when decanting the liquid.

The manufacture of these transport workhorses was complex. First, the main body of the pot was formed, leaving a small hole in the base to allow for quicker drying. Then the neck and rim, created separately, were joined to the main body and the hole at the bottom was closed up. The two side handles were added last. Pot production took place on an industrial scale. Along the banks of the Guadalquivir (Spain's longest and most important river) are the remains of at least a hundred pottery workshops, some with rows of kilns that would have occupied dozens of workers in the production of shipping amphorae. Since land transport was expensive, it made sense to manufacture these heavy pots near to rivers or coasts, and the Baetis, as the Guadalquivir was then known, was a vital artery in the olive trade, running from Córdoba out into the Atlantic Ocean.

Baetica's mountainous land and the hot, dry climate provided perfect conditions for nurturing the source of its wealth, the olive tree. Pliny the Elder, the first-century philosopher and author of the famous *Natural History*, remarked on Baetica's 'peculiar brilliance of vegetation'. However, as well as abundant supply, Baetica had that other vital ingredient in wealth creation – a stable market. The olive oil supply chain combined state-controlled production and free-market economics, all driven by demand from Rome. Rations of olive oil from Baetica are thought by some to have been distributed through the *annona*, an official agency that also distributed grain. Sensibly enough, the imperial authorities reckoned that well-fed people made for happy citizens. But they allowed private merchants to participate in provision of the supply. In addition, the Roman army needed feeding, and Baetica was the chief source of olive oil consumed by the legions stationed in outposts such as Germany and Britain.

Drawn by the empire's appetite, millions of jars of olive oil made their way from Hispania to Rome, departing from places such as Córdoba and sailing southwards via the Balearic Islands. Some went along the North African shore and across to Italy. Amphorae made other journeys, too, finding their way to markets in Italy, France, Britain and even India. The Romans, the great road builders, were less comfortable on water. Even when they gained control of 'Mare Nostrum' (their name for the Mediterranean), the Romans preferred navigation techniques and trade routes that did not take them too far from land. Yet water transport was a crucial part of the economic equation, since hauling goods by land was extremely expensive. Their ships were simple wooden affairs. Rather than using overlapping planks to create the hull, these vessels were constructed using the tongue-and-groove system, where the edge of each plank slots into that of its neighbour. Deck, mast and internal ribs were then built and the lower part of the hull was sheathed in lead to protect it from wood-boring sea worms.

Conveyed in these vessels, oil from Baetica ended up at the port of Ostia, on the Tyrrhenian Sea off Italy's west coast, where, after a short layover in a warehouse, it was loaded on to barges that could navigate the Tiber and unloaded at port installations along the river's banks in southern Rome. As well as receiving olive oil shipments, Ostia was also where huge shiploads of grain arrived from Egypt to be ground into flour, baked into bread and sent on the barges up the Tiber into Rome. Echoes of this mighty trading post can be found today at the remains of the Roman port, now known as Ostia Antica, about forty minutes' drive south-west of Rome. The mill-bakeries with their giant grindstones and ovens survive and, wandering through the paved streets and impressive public buildings of this magnificent archaeological site, it is easy to picture a thriving maritime city of warehouses, shops, lavish apartment blocks, grand theatres and elegant villas. Here, the Romans enjoyed themselves at luxurious baths as well as at brothels and bars.

If the remains at Ostia conjure up images of the lifestyle of wealthy Romans, what funded such decadence was transportation and international commerce. From within small booths at the Square of the Guilds, a sort of trading floor, vessel owners from different parts of the world haggled over freight rates and merchants negotiated business deals and supply contracts. Among the guilds were the *mercatores olearii* (oil merchants). On one side of the square, a floor mosaic provides a visual reminder of their trade – a slave aboard a vessel carrying a large amphora on his shoulders. While oil merchants talked money in the Square of the Guilds, these slaves were out on the docks, their backs glistening with beads of sweat as they unloaded pot after pot of olive oil beneath the fiery Italian sun. And those pots were extremely heavy. An amphora weighed about 66 pounds when empty. When full, carrying about 6 gallons, it weighed more than double that. For the slaves, then, the amphora was a heavy burden. For the men and donkeys lugging broken shards up the slopes of Monte Testaccio, the amphora was also a chore. But for the olive merchants of Baetica, the amphora was something quite different – an extraordinarily efficient ceramic vessel at the heart of an international trade that thrived many centuries before the word 'globalisation' had been coined.

These days, olive oil does not arrive at the point of purchase (the POP, in the retailer's jargon) in a ceramic pot. The olive oil we buy comes in glass – in sleek designer bottles. Lined up on the shelves of shops and supermarkets, these transparent vessels of golden-green liquid have a jewel-like allure, evoking amber or polished jade. The more expensive varieties are even corked and sealed with wax, giving them a grandeur associated with fine wine. Lit from behind, the bottles have a marvellous luminosity. Their seductive quality is all about aesthetics and image. Taste does not enter into the equation, at least not until the bottle is safely home (not even the faintest odour of the contents will be released until the seal is broken). Then there are the labels. The more traditional among them show beautifully crafted images of rustic farm labourers and women in nineteenth-century dress. Art deco flourishes, impressive-looking crests and classical ornamentation embellish these pictures. An entire industry revolves around designing olive oil bottles and coming up with images that will help us conjure up a whiff of the Mediterranean, whether we are wandering round in a shopping mall in Dallas or popping into the local supermarket on a rainy day in Leeds.

When it comes to marketing olive oil, the Italians are the experts. The Bertolli website provides a hint of this mastery. Like shavings of fresh Parmesan, romantic images of *la dolce vita* scatter gently across the computer screen as the presentation downloads. In this particular section of cyberspace, everything evokes a bygone era – an Italy seen through rose-tinted spectacles, where Mama cooks the pasta, old men with faces weathered as an olive tree's bark tend the groves and dinner is served on a terrace overlooking a sun-baked Tuscan landscape. The website tells us that back in 1865 – just seven years before Bruzza and Dressel were to make their first excavations on Monte Testaccio – Francesco Bertolli and his wife established a shop at the front of their house in Lucca, in the Tuscan heart of Italy's olive-growing region. That building on Piazza San Donanto (it is still standing, we are told) is where the business began. By buying Bertolli olive oil, the website suggests, we can all savour 'life the way Italians do'.

It is an appealing thought. That is, until you pick up a bottle of olive oil at the local supermarket, thousands of miles from that Tuscan landscape, and find that while, on many brands, the label proclaims it has been 'imported from Italy', the golden liquid within is often (as revealed in the small print on the back) a blend of oils from Italy, Greece, Spain and Tunisia. For all the clever associations conveyed by the packaging, the reality is that what we're buying is not entirely Italian. Much of it has been picked and pressed elsewhere and transported across Europe before being blended, bottled, labelled in Italy and sent out on its way again.

Much of the oil in the bottles of Italian companies will have started its journey in Spain. In this respect, little has changed since Roman times. Spain, not Italy, is still the world's most important source of olive oil. Spain is the origin for more than half the olive oil produced by European Union countries (Italy's share is about thirty per cent) and the country's olive groves represent more than twenty-five per cent of the planet's olive-growing surface area. The International Olive Oil Council, an organisation chartered by the United Nations to regulate the world's olive oil trading, operates from Madrid, not Rome.

Most of Spain's oil still comes from Andalusia. Here, olive trees creep over every inch of the terrain, peppering rocky outcrops or marching aggressively in ranks over burnt-out hills. Architecturally, much of this region echoes with memories of the Moors. Yet evidence of the Roman era also survives.

Magnificent Roman-built bridges still span the rivers at Córdoba, Mérida and Alcántara and the ruins of Roman towns such as Baelo Claudia, once a fish-salting centre, still stand. However, the strongest link between modern Andalusia and Roman Baetica remains the olive. Andalusia is home to more than 200 million olive trees, about a quarter of which grow in Jaén, a small province that is responsible for an astonishing twenty-two per cent of the world's olive oil.

As in Roman times, much of this oil travels great distances. Today, about sixty per cent of Spain's massive output heads to Italy. Much of it ends up in processing plants where it is blended with other oils, packed into bottles, and sent out again. This leads to some odd-looking figures in the trade statistics. When it comes to production, Italian output is half that of the Spanish co-operatives and most of the European Union's olive oil imports head towards Italian shores. Yet Italy accounts for sixty per cent of Europe's olive oil exports – almost double that of Spain. Behind these figures is a massive shunting of oil from one place to another. Arriving in bulk, the oil has a short layover in Italy and is sent out on its way again, dressed up in a smart bottle and adorned with an ornate Italian-style label. Not everyone is happy about this. Spanish producers grumble about export subsidies for Italian companies, as well as the fact that their oil ends up in bottles that look Italian. Manuel Lopez, the export manager at Hojiblanca, an olive oil co-operative based in Andalusia, concedes that the Italians have done a good job of selling oil to the world. 'They have always had very good marketing people,' he says. 'And Spain has been sleeping in this respect. We need to get back into the market again.'

Spanish producers are trying to do just that. They know

Spain is the pre-eminent source of olive oil – now they want the rest of the world to know it. The campaign got going in the 1990s, when the Spanish government provided funds to help local olive oil companies market their oil overseas. Since then, Spanish co-operatives have been designing smart labels for their bottles and are trying to sell more of their oil this way, directly to consumers, rather than having it disappear into the giant tankers heading for Italy. Marketing efforts have also helped. In the USA, now one of the world's biggest markets for olive oil, supermarkets run frequent promotions and tasting events, and distribute leaflets singing the praises of Spanish olive oil. One Spanish company has even enlisted Hollywood. Andalusia-born movie star Antonio Banderas has taken a stake in Hojiblanca, Spain's largest olive oil co-operative, and has been promoting the company's oil.

The trouble is, names such as 'Hojiblanca' are hard enough for some Europeans to pronounce, let alone for people in countries such as China, Japan or Taiwan. Moreover, consumers in these markets hold great potential for Spanish exporters as they do not yet associate olive oil with the Italians alone. Yet Spain's olive oil companies have tended to promote the region where their oil is produced, rather than pushing the idea that it is 'Made in Spain'. This is something the Spanish government wants to change. It has set aside a chunk of the subsidies it hands out to producers for a marketing programme it hopes will lead to the evolution of a generic 'España' brand for Spanish olive oil. The idea is that the world will start to associate olive oil with Spain, as well as Italy.

As transport makes it cheap and easy to fill supermarkets with food and drink from all over the world, the geographical origins of the things we consume are, like olive oil, not always where we think they are. Foods are showing up in the most unlikely places. Who would have predicted that China would become a source of foie gras and truffles? Who knew that Maharashtra state in India would start producing some decent wines or that an enterprising farmer in Punjab would nurture a fledgling business growing 'Florida' oranges? What remains to be seen is whether, once these foods arrive in foreign markets, the labels and advertising campaigns will disguise their origins. Odds are, Chinese foie gras will not come in packaging decorated with pictures of the factories where tens of thousands of geese are being force-fed.

While labelling rules are becoming stricter, we have certainly been tricked in the past. Early Arab traders kept the source of spices a secret to preserve their monopoly on the valuable commodities, as did the Portuguese, whose king decreed in 1504 that anyone publishing details of the origin of the spices Vasco da Gama had discovered on the west coast of India would be executed. Today, subterfuge is still at work. Until recently, shops routinely sold 'Scottish smoked salmon' that could have been fished anywhere from Norway to Chile (today, only fish from Scotland can be labelled 'Scottish'). The many foods branded as 'farmhouse' – anything from cheese to jam – have usually come from a large factory and not a rustic home nestling on a bucolic stretch of farmland. Moreover, sometimes places that are entirely fictional are created to add to the appeal of a food. British chain Marks & Spencer recently introduced 'Lochmuir salmon', despite the fact that Lochmuir cannot be found on any map.

Our desire to associate food with certain parts of the world often drives our choices of what to buy – and marketers know this. Fiji, a brand of bottled water, makes the most of its location (it even puts a helpful map on the back of some of its bottles). With tropical flowers on its label, its website reminds shoppers that the water originated in a place far from pollution, acid rain and industrial waste and was drawn from an artesian aquifer, located at the edge of a primitive rainforest, 1,500 miles away from the nearest continent. 'There's no question about it: Fiji is far away,' we are told. 'But when it comes to drinking water, "remote" happens to be very, very good.'

Labels such as these tap into the fact that the food we buy often allows us to travel vicariously. 'Remember this,' the bottled water company tells us. 'We saved you a trip to Fiji.' What we create in our kitchens can transport us to anywhere from Casablanca to Beijing. We happily dream of Italy as we slice up our mozzarella (which could have come from anywhere, although *mozzarella di bufala campana* can be made only in Campania), drizzle some olive oil 'imported from Italy' (that might have come from South Africa, Egypt, Turkey, Israel, California, Greece, Spain, Morocco or Tunisia) over tomatoes (that could have come from Spain, Mexico, Chile or Florida) and throw in a few leaves of basil (from the plant growing on the window sill) *et voilà*! Or rather '*ecco*!' – you have Italy on a plate. Well, sort of.

Food helps us engage in a bit of virtual travel. Though we know that these are imaginary journeys, part of us would still like to feel that these things really came from the places with which they have historically been associated. So if they start exporting their Florida oranges, will the farmers of Punjab really be brave enough to brand them 'Punjabi oranges'? If not, marketers will work hard to present us with the images we expect to see – little white lies that help sell the product. To satisfy labelling legislation, the information will be there in the small print, but labels take time and energy to scrutinise and most of us want to get in and out of the supermarket as quickly as possible. And in the end – as those working in the news media will readily admit – it is images, not words that tend to stick.

Of course, some foods slip their geographical moorings altogether, with their places of origin devolving into generic names. Surely no one can be under the illusion that the Brussels sprouts they tuck into at Christmas actually grew on a field in Belgium or that the meat in Maryland crab cakes is from creatures that were fished off the small state on the US east coast. As we crumble some Demerara sugar into our baking, how many of us are thinking of the former colony in the Caribbean, part of British Guyana, that first grew the cane needed to produce that sugar? 'French' beans could have been grown anywhere from Kenya to California. Few people conjure up images of the famous caves of Cheddar Gorge when they are savouring a slice of the tangy cheese originally made nearby. And rightly so, it turns out – one of the largest suppliers of cheddar cheese is not rural Somerset, but a giant industrial production plant constructed on a 3,000-acre campus in New Mexico. Some of that cheese may be among the thousands of tonnes of cheddar bought and sold every day by a small group of commodity traders who gather every day on the trading floor of the Chicago Mercantile Exchange and gesture wildly as the market opens. It is all a far cry from the English village where cheddar cheese making began, some say, after a milkmaid accidentally left a pail of milk in one of the caves and returned later to find it transformed into a delicious creamy substance.

Today, roquefort, caerphilly, cheddar and emmenthal are all cheeses, not places. Darjeeling is a tea, not an Indian town on the lower slopes of the Himalayas. Champagne is a sparkling wine, not a region in France. However, the use of place names – or names closely associated with certain places – to describe food has generated frequent legal battles, particularly from countries whose signature foods are being replicated elsewhere. One of the biggest fights was settled recently after a two-decade wrangle between Greece and other producers of feta cheese. The Greeks claimed the sole right to sell the pale, crumbly sheep's milk cheese under the name 'feta'. Others – including Germany and Denmark – disagreed. The European Union ruled in favour of the Greeks and now, cheese sold in Europe can be branded as feta only if it has come from certain regions of Greece.

The feta battle is just the start of it. Across Europe, producers of everything from Parmigiano Reggiano and Prosciutto di Parma to Melton Mowbray pork pies, Newcastle brown ale and Arbroath smokies (haddock smoked over hardwood) have embarked on crusades to win back the names of their products. The European Union has generally been sympathetic. It has devised what is known as a geographical indication system – rather like the *appellation d'origine contrôlée* certification system that since the 1930s has governed French wines. True to form, the bureaucrats in Brussels (who curiously have not shown any interest in reclaiming their sprouts) have come up with a series of acronyms for the system – PDOs, PGIs and TSGs. Here is how they work: something with a protected designation of origin (PDO) must be produced in a specific geographical area using a recognised method; to register a protected geographical indication (PGI) at least one stage in the food's production must take place in the designated region or country; and a food given the tradition speciality guaranteed (TSG) designation must stick to the proper method of production or preparation. It is a nicely ordered system.

But like any food fight, this one is messy. The rules apply only to foods sold inside Europe. The EU has no authority to prevent American or Mexican cheese producers naming their products 'feta' or 'parmesan' and selling them in America or Mexico.

What is more, non-European food companies claim that the new rules are simply another trade barrier keeping their products out of European markets. In line with a World Trade Organisation ruling, Brussels has now allowed companies based outside the EU to sign up to the geographical indication system. However, it is still easy to see why a Danish feta producer, having had to rebrand a perfectly successful cheese, might not be happy to see that American companies selling their cheese outside Europe can continue to use the name. The battles are set to continue, fought out at future rounds of international trade negotiations.

Some Europeans have successfully protected their names overseas without the assistance of global trade rules. The champagne and wine growers of France have won agreements from other countries not to use names such as Burgundy and Champagne on their wines. Even Yves Saint Laurent, the perfume maker, was in the mid-1990s forced to alter the name of one of its fragrances from 'Champagne' to 'Yvresse' after a long legal battle with French champagne growers. One enterprising South African, Charles Back, has fought back – with a smile on his face. He named his wines Goats Do Roam, and soon added a Bored Doe and a Goat Door (although

Fairview Estate, Back's vineyard, claims the names are not in fact poking fun at the Côtes du Rhône and other French growers but arose when some goats from the Fairview herd escaped their paddock and wandered among the vines, chomping on ripe grapes).

Back's sense of humour is unusual. For others, like the Melton Mowbray Pork Pie Association, the battle over food names is deadly serious. The association reckons the market for its pork pies is worth £50 million a year – revenue it wants to keep for Melton Mowbray, the ancient market town where the pies are made. Yet producers elsewhere lay claim to the name, saying they have been making Melton Mowbray pork pies for years. These sorts of battles are likely to proliferate. In a world where transporting food is cheap and efficient, national or regional laws simply cannot keep up with the increasingly international flavours of our moveable feasts.

The Romans did not worry about any of this. For the Baetican estate owners, the amphora helped them trade their produce in far-off markets. No subterfuge was deployed as they sold their olive oil. Hispania was a proud part of the Roman Empire – and supply and demand was what kept that empire together. The imposition of Roman rule was therefore good news. It instantly created a lucrative international market for Baetican merchants. In any case, the Romans had created a multicultural mix in Europe that was unmatched in the rest of the ancient world. With it came a truly global approach to eating. Roman citizens everywhere from Germany to Egypt were accustomed to the fact that what they ate had been hauled in from all over the empire. In Rome itself, there was grain from Egypt and Africa, wine from Spain, France, Greece and Sicily, preserved fruits from Syria, walnuts from Persia and rare spices from China, India, Arabia and Africa. Southern Indian pepper dating from the first century has been found in Germany, then known as Germania. Even fresh produce travelled some distance to reach consumers – apples and pears were conveyed by road down to Rome from Picenum, a northern district between the Apennines and the Adriatic. And the pre-eminent source of Rome's olive oil was, of course, hundreds of miles away.

Today, the oil travels in vessels that look quite different from those used by Roman shippers. Giant tankers made of stainless steel now ply the roads across Europe, each carrying many tonnes of olive oil. The tankers must be kept airtight as oxygen hastens the rate at which the oil turns rancid. Temperature extremes must be countered, too. A cold snap could solidify the oil, making it impossible to pump out. At $10°$ Celsius, the oil separates. At $6°$, it becomes semi-solid in consistency and at $0°$ it turns into something resembling butter. Olive oil tankers use heating coils on the walls to keep the contents sufficiently sloppy during the voyage. At the end of their journey the oil is pumped out through large pipes into vats before being bottled at the processing plant and re-exported. It is not an image that immediately springs to mind as we sprinkle a few drops over our salad.

Transport technology may have altered beyond recognition since Roman times, but in our appetite for foods that have travelled from all over the world, are we really so different from the Romans? Perhaps not. In fact, what makes this remote civilisation so compelling is the fact that, despite its distance from us in time, the lifestyle of the Romans feels oddly familiar, particularly when it comes to eating. Back in Ostia's ancient ruins, one can almost smell the bread rising in the ovens at the bakery – bread made from Egyptian flour ground in one of its millstones. At what would have been a sort of tavern in one section of the city, it is easy to imagine savouring a glass of wine from a consignment that might have arrived from Greece and had been unloaded on the docks at Ostia that morning – Greek wine accompanied by a selection of preserved fruits from Syria or a bowl of nuts from Anatolia (a region roughly corresponding to modern-day Turkey). Then for dinner, perhaps a tuna steak seasoned with black pepper from India and a dish of roasted vegetables baked in olive oil – a golden liquid from southern Spain that was conveyed to Rome in an amphora. Empty of its contents and broken into fragments, that pot now lies in a gargantuan pile of shards beneath the grassy slopes of Monte Testaccio.

2

The Business of Bones

The Norwegian salmon pays a visit to China

Salmon: Fish of northern waters with delicate pinkish flesh

Origin: Freshwater streams, estuaries and hatcheries

Etymology: Possibly from Latin *salire,* to leap

Legends: Celtic lore has it that the salmon gained its wisdom by eating nine hazelnuts that fell into a pool from the Tree of Knowledge. Finn, pupil of an old druid, became a great Celtic leader after popping a blister on a salmon as it cooked. Sucking his burnt thumb, he inadvertently acquired the fish's knowledge

*F*OR ORDINARY HUMANS, the extra-ordinary migration of the salmon – Scottish or otherwise – is difficult to imagine. Born in freshwater streams, the young fish, known as a smolt, spends the first few months of its life in the river before embarking on a monumental voyage to the ocean. It remains at sea for up to three years, feeding, fattening – and travelling thousands of miles in the process. Once the salmon reaches sexual maturity, it begins the long voyage back to its roots to procreate. In a most mysterious journey, the salmon returns not only to the river of its birth, but to the exact same tributary of that river, battling upstream most of the way. Pictures of gleaming fish flying through the air as they defy gravity to negotiate the rapids of large rivers are some of the most poignant images in the portfolios of wildlife photographers.

The immense physical struggle of this journey epitomises a strange kind of heroism – because the fish is not only returning to its birthplace to start new life. It is also travelling to the place where it will die. As it re-enters the river after its three-year sojourn in salt water, the darkening of the fish's skin is the first sign of the onset of its decline. Even as it leaps athletically up waterfalls of more than 12 feet in height, the salmon is dying. On reaching its birthplace, having spawned its eggs and given life to the next generation, it starts to weaken. Eventually it will be washed downstream to end its life as an easy lunch for a bear or an eagle.

Perhaps the most remarkable aspect of this journey is what powers the homing instinct guiding the salmon throughout this long and arduous voyage – its nose. The technique is not fully understood. However, studies have come up with the conclusion that the salmon's sense of smell is crucial to its ability to find its way home. The first study was conducted in 1951. In the experiment, two scientists, Warren Wisby and Arthur Hasler, blocked the noses of a large group of salmon and found that their return journeys were far less accurate than groups they monitored without the nose plugs. Studies in which the salmon had their olfactory nerves cut have produced similar results.

The evidence emerging from these studies is compelling. Yet anyone wondering how the remembered smell of a river can, by itself, guide the salmon back from distant oceans is not alone. Some have speculated, for example, that it is the earth's magnetic field that guides the fish. The idea is that once the young salmon enters salt water on its voyage out to sea, a memory of that particular latitude and longitude is imprinted on the fish as chemical and hormonal changes take place in its body. This, it is thought, could create a sort of biological compass that is set the moment the fish enters the ocean. Alternatively, the fish itself might act as a conductor, able to detect a magnetic field whenever it passes across one.

The methods of homing may remain shrouded in mystery, but the routes taken by salmon during their lifetime are about to become clearer. As part of the Pacific Ocean Shelf Tracking project, scientists are engaged in an unprecedented effort to monitor the lives of Pacific salmon. The idea behind the project (dubbed 'fish with chips' by some) is to insert tiny computer processors into young fish and then track their progress across rivers and oceans. Employing technology rather like the systems used to extract tolls from drivers on highways, the scientists implant small electronic tags in them that are scanned whenever they pass across tracking devices installed on the ocean floor. The studies are yielding surprising results. Among the project's findings so far is the fact that Chinook salmon smolts swim from the Columbia and Snake Rivers up to Canada and beyond, covering up to 16 miles a day. Calculated as body lengths per second, that would be the equivalent of a human swimming more than 160 miles a day – fast enough to circumnavigate the equator in 150 days. As well as achieving impressive speeds, migrating fish cover vast distances. In its trans-Pacific migration, a tagged bluefin tuna was found to have covered an amazing 25,000 miles – a distance greater than the earth's circumference.

If the mileage clocked up by these fish sounds impressive, it is nothing compared to the journeys some of them take after their death. In the case of salmon, it is all because of their pin bones – dozens of tiny bones not connected to the rest of the fish's skeleton that cannot be dealt with by filleting machines. Pin bones must be extracted by hand using tweezers or small pliers. It is a laborious process that, if it is carried out in North America or Europe, is a costly business. Not in China, though, with its low wages and high productivity. So companies send the fish to the Far East, where they are defrosted, filleted, refrozen and shipped either to Japan, or back to western markets (the refreezing is safe, although there can be loss of quality because of the moisture generated by defrosting).

A dramatic increase in the amount of fish travelling to China reflects the growth in this method of filleting. By 2002, the country had become the world's eighth biggest fish importer – five years previously, it had not even made it into the top fifteen. To be sure, some of the fish stays in the country to be consumed there. However, a substantial proportion of the fish imports can be accounted for by the arrival of what those in the business call 'raw material' – not only salmon but also cod, pollock, crab, flounder, yellow-fin sole and all sorts of other fish. These creatures are short-staying guests on Chinese shores until, deboned and cut into neat pieces, they embark on their journey back west.

Here is a typical journey for a Norwegian salmon destined for sale in a supermarket in America or Europe. Once harvested, the fish is frozen and packed into boxes that are loaded on to a small feeder vessel, most probably at Alesund, an important fishing harbour and a place whose natural beauty and famous art nouveau architecture make it one of the highlights of a tour of the Norwegian fjords. From this picturesque starting point, the fish sails to Rotterdam or Hamburg, where it will change ships and end up on a large international container vessel bound for the Far East, travelling at a temperature of −23° Celsius all the way. Thousands of miles and about a month later, the fish arrives in China, often ending up in Qingdao, a large port city on the tip of the Shandong Peninsula of China's north-east coast that is home to several hundred fish-processing centres. Qingdao is an odd place. Ugly skyscrapers rise up next to ornamental pagodas above an incongruous-looking collection of low-slung Bavarian-style houses and the occasional Gothic church. These European oddities date from the late nineteenth century, when the city was annexed by Germany, which used it as a naval base and established breweries in the city (Tsingtao is now China's favourite beer). The Germans left long ago. Once again, however, foreigners are helping shape Qingdao's industrial landscape – and much of the activity is focused on fish.

After being unloaded from the vessel the 'raw material' is trucked to a fish-processing centre on an industrial park. At this smart new facility with vast cold storage facilities, the salmon is defrosted and moved out to the factory floor. In an immaculately clean industrial space (Chinese managers know they will lose business if they do not comply with strict European and American hygiene regulations) are ranks of tables, each with dozens of brightly coloured plastic trays on top of them. Standing at the tables, dressed in white coats and caps and wearing latex gloves and cotton masks, are hundreds of factory workers – most of them young women from rural villages around China. Using nimble fingers and small scalpels, they swiftly skin the salmon, remove its bones and cut it into the exact portions specified by a western supermarket chain on the other side of the world. It is an efficient assembly line – although what is going on here is more like 'de-assembly' than assembly. Once the fish is filleted and in pieces, it is refrozen, packed back on to a ship and sent back to Europe. How can such an insane journey be economically viable?

Enter the shipping container. For half a century or so, this steel box has been transforming world trade. The swinging sixties may have ushered in social, sexual and musical revolutions, but down on the docks another revolution was gathering momentum: 'intermodalism'. It does not sound tremendously exciting. However, intermodalism is an astoundingly efficient system whereby one-size containers can be shifted seamlessly from ships to trucks and trains. Throughout the 1970s and 1980s, this system quietly spread to every corner of the global transport network, dramatically reducing shipping costs and paving the way for a massive growth in world trade.

The container helped transform the global economy, but this particular revolution has escaped our attention. While few people notice containers or think about what is inside them, more than ninety per cent of global trade travels in these metal boxes, with more than 200 million containers moving across the oceans each year. The amount of stuff carried around in these boxes every day is mind-boggling. A single 20-foot container would hold about 48,000 bananas. Today's vessels each carry about 8,000 of these containers, which if placed end to end on top of each other would reach higher than 160 Eiffel Towers. Within these unassuming boxes lies the stuff of our lives – the glasses we drink from, the beds we sleep on, the computers we curse and the juicy mangoes into which we sink our teeth. Almost everything we touch, from door handles, cutlery and furniture to bicycles and television sets (and, of course, frozen fish), has at some point been inside a container.

The brains behind this remarkable box was a North Carolina truck driver born in 1913, Malcom McLean. Worried about the competition from ships to his trucking business, McLean came up with the idea of simply driving loaded trailers on to vessels – his own vessels. As his ideas developed, he realised that a far more efficient use of space would be to do away with the wheels of the trailers and simply use the trailer bodies. On 26 April 1956 he put his idea into practice. That day, what one reporter described as an 'old bucket of bolts' set sail from the port of Newark, New Jersey, and headed down the east coast for Houston, Texas. The vessel – a converted Second World War tanker – was a curious-looking craft. But with a reinforced deck carrying fifty-eight metal boxes, Malcom McLean's *Ideal-X* initiated the first-ever scheduled container ship service.

McLean's invention was to end break-bulk shipping, a

laborious system whereby each piece of cargo had to be separately loaded, packed, arranged and unloaded – a practice that had altered little in centuries. Suddenly, instead of having to deal with bulky, breakable, irregular objects, freight handlers simply slotted together tough metal boxes of standard sizes. With the old process consigned to history, the container has allowed goods to be transported cheaply, speedily and in previously unimaginable quantities.

What is more, the journeys of these steel boxes are far easier to keep track of than the old break-bulk shipments. Each container has a unique code on its door, rather like a car licence plate. It is a collection of letters and numbers that gives each box – in every other respect identical to millions of others – its own identity. The code denotes the owner of the container and what types of product it is able to carry. Some of the numbers link the individual container within the shipper's fleet; others describe details such as the dimensions of the container, its weight when empty and its weight when full. For ship's captains, crews, coastguards, dock supervisors, customs officers and warehouse managers the code is an instant guide to what is in a container, where it came from and where it is going – not so different, after all, from the stamps, marks and ink inscriptions found on the ancient Roman amphorae used to transport olive oil. Today, however, a new agenda is behind added layers of identification for the container – security. The threat of a terrorist putting a 'bomb in a box' has focused government attention on the thousands of containers arriving at the world's ports every day and prompted technology experts to come up with ever more sophisticated scanning and tracking devices. In the post-9/11 world, radio frequency identification technology, intrusion detection systems, long-range ground surveillance radars and thermal imaging technologies have all entered intermodalism's lexicon.

McLean was not thinking of terrorism when he dreamed up his system of moveable metal boxes. Speed and efficiency was his chief concern, and he reckoned he had hit upon a winning formula. The idea was slow to catch on, however. Port operators and ship owners were reluctant to commit to expensive new equipment, vessels and port infrastructure. Down on the docks, the unions saw their members' jobs under threat and resisted the proposed changes. Then in the 1970s, the global oil crisis meant fuel prices rocketed to such levels that the container was unable to make much of a dent in shipping rates, despite its efficiency. Eventually, however, the logic of the system proved too attractive to ignore. After all, getting men to hoist cargo on and off ships by hand in individual bags or boxes accounted for about half the total cost of shifting goods between ports. The new system cut labour costs by about ninety per cent. Moreover, containers did not need to be kept in expensive warehouses but could simply be stacked up in yards. Intermodalism was starting to look a lot more exciting.

While it was busy reshaping global trade, the container was also transforming cargo-carrying vessels, turning them into gigantic low-slung craft, capable of carrying several warehouses'-worth of goods. The world's biggest is a ship called the *Emma Maersk*. She is immense – almost 1,300 feet long, with a beam (the width at the widest part) of more than 180 feet. Her engine weighs 2,300 tonnes, her propeller weighs 130 tonnes and she has twenty-one storeys between the bridge and the engine room. Seen alongside the docks, she appears more like a floating industrial plant than a ship. Out on the ocean, she could hardly look more different from the ancient Roman vessels packed with amphorae or the galleons, frigates and corvettes that were her maritime predecessors. No longer is this a vessel crawling with slaves or sailors and fitted out with complex rigging. The *Emma Maersk* can be operated with a team of just thirteen people and a sophisticated computer system. Gone, too, is the idea of stowing everything in the hold. Now the deck acts as a vast tray on which the boxes are stacked high. Slicing her way efficiently through the water at about 26 knots, the *Emma Maersk* carries an astonishing 11,000 20-foot containers – the equivalent of a train 44 miles long. That means if there were a freakishly big rush on those bananas, she could transport more than 500 million of them in a single voyage.

Unloading cargo ships, which once took large teams of men a couple of days, is now done in a matter of hours by a handful of workers using extremely heavy-duty machinery. The large steel packages these workers handle also give the objects within them unprecedented protection from damage and pilfering, thereby lowering insurance costs. Theft has always been a problem on the docks. In New York Harbour, things were once so bad that, according to one story, an Italian shoemaker was forced to devise a system to prevent his goods from being stolen – he would ship his shoes to New York in two separate consignments, one for the left shoes, another for the right. When in the old days whisky travelled in cardboard boxes, dockworkers at both ends of the journey found it easy to whip out a few bottles to take home. Now the bottles are stuffed into a steel container in the distillery in Glasgow and not let out again until they have reached the warehouse at the other end.

Some regret the passing of the old days, the colourful activity on the docks and the lengthy stays in port, giving the crews time to enjoy themselves in town. Nevertheless, there is something thrilling about watching a large modern container vessel sail into a harbour. Dwarfing the tugs and other small craft around her, she looms up on the horizon while on land everyone stands in anticipation as she slides towards the docks. Gantry cranes straddle the quayside like gigantic stick insects, waiting to move in on their prey. The moment of calm soon vanishes. The second she comes to a standstill, an army of moving machines emerges from behind warehouses ready for the attack. Harried supervisors in smart white boiler suits bark instructions down walkie-talkies and shipping agents squint at dog-eared papers on clipboards. Everyone is in high gear – and so they should be. A vessel makes money while moving but loses it when docked. Speedy turnaround is essential. As the unloading operation gets going, the gantries are roused from their slumber to lurch toward the deck and lever the contents out into the brilliant sunshine. Chunky ten-wheeled cranes roll up and down the quayside making light work of stacking heavy cargo. Nifty forklift trucks wheel around at speed, shifting, lifting, reversing, shunting – nothing rests on solid ground for a moment longer than necessary.

The forklifts and gantries are, however, just one part in the complex chain that is intermodalism. Waiting in the wings as the unloading continues are trains and trucks. Soon the containers will be transferred on to their trailers and sped overland by rail or road to factories and warehouses. This is the real beauty of McLean's invention. Others before him had used metal boxes for transporting goods, but no one had considered handling freight through an integrated system of transport – a system in which every piece of equipment, from crane and forklifts to trucks and vessels, was designed to handle a single, uniform piece of cargo: a moveable metal box.

Once upon a time, shifting the goods took brawn. In the nineteenth century, poor workers from London's East End queued up every morning for casual work and, if they were lucky, got to spend a day heaving sacks from a vessel to nearby warehouses. Henry Mayhew in his 1861 book, *London Labour and the London Poor*, described the early morning scene, when large crowds of men huddled round the London Dock gates, hoping for a day's work. 'Presently,' wrote Mayhew, 'you know, by the stream pouring through the gates and the rush toward particular spots, that the "calling foremen" have made their appearance. Then begins the scuffling and scrambling forth of countless hands high in the air, to catch the eye of him whose voice may give them work.'

That work was gruelling. At West India Quay in the heart of the docklands, the quayside was nicknamed Blood Alley because the sugar in the sacks carried into the warehouses scratched the dockworkers' backs until they bled. After the day's labour, workers would head to the nearest bar and drink away their earnings. In the 1860s, in *The Uncommercial Traveller*, Charles Dickens wrote:

> Down by the Docks, the seamen roam in mid-street and
> mid-day, their pockets inside out and their heads no better
> . . . Down by the Docks, anybody drunk will quarrel with
> anybody drunk or sober, and everybody else will have a hand
> in it, and on the shortest notice you may revolve in a
> whirlpool of red shirts, shabby beards, wild heads of hair,

bare tattooed arms, Britannia's daughters, malice, mud, maundering, and madness.

Things were no better on the other side of the Atlantic. Organised crime was rife and port managers put together temporary gangs of men that they picked out at the morning 'shape-up'. On the New York waterfront, the system bred cronyism and corruption. 'Men are hired as if they were beasts of burden, part of the slave market of a pagan era,' Father John Corridan, a passionate advocate for labour reform, told the author of *The Waterfront Priest*, written in 1955. Perhaps the most striking depiction of pre-container dock work appears in the 1954 movie *On the Waterfront*, which was based on features written for the *New York Sun* exposing corruption on the New York docks. The film captures the days when stevedores (the word comes from the Spanish *estibar*, which means 'to stow' or 'to pack') did all the heavy lifting. Playing the muscle-bound dockworker Terry Malloy, Marlon Brando famously declares: 'A man's gotta do what a man's gotta do.' Back then, that meant shifting boxes, crates and sacks with the aid of primitive equipment: booms, nets, chunky metal hooks – and brute force.

Fast-forward to the port of Hamburg today and things are very different. The place is empty of dockworkers. Gantry cranes, forklift trucks and other vehicles are operated remotely from control offices overlooking the container yards. The range and sophistication of the machines that containers have spawned is astonishing – everything from self-loading container trailers to reach stackers, terminal tractors, trailer winches and automatic stacking cranes. In Brisbane, another high-tech port, the automated straddle carriers built by

Kalmar, one of the world's biggest suppliers of container handling equipment, are in charge of moving things around. Fitted with motion control sensors, radar and navigation systems, these oversized robots roam freely around the terminal, handling up to thirty containers an hour, twenty-four hours a day, throughout the year in all weather conditions without the intervention of humans. In this world, size matters too. The biggest gantry cranes are almost 400 feet tall, weigh up to 1,500 tonnes and could lift weight equivalent to thirty elephants. The newest ones can reach across twenty-two containers to grab the one they are after. Standing tall like triumphal arches that have been painted in primary colours, these are the industrial monuments of global commerce.

A decade after the release of *On the Waterfront*, machines were starting to put labourers like Malloy out of a job. By 1985, the number of dockworkers employed in New York Harbour had plummeted from more than 30,000 in 1960 to just 8,500. By the 1980s, much of the London docklands lay derelict as port operations moved outside the city. As similar developments took place at ports around the world, some of the most bitterly fought union battles of the last century have been set on the docks.

The container has arguably been one of history's greatest agents of change – and not just for the 'wharfies' who lost their jobs. Containers brought consumers new and exciting goods from remote places. When western companies found low-cost suppliers in countries such as Taiwan, South Korea, Singapore and Hong Kong, the container made it possible to transport the goods at a cost that was tiny compared to the sale value of those goods. Americans and Europeans could acquire cheap television sets and washing machines. Rattan furniture became fashionable. People bought things they previously would have never dreamed of possessing as the container helped foster the consumer revolution. Today, virtually everything we buy has come from another part of the world.

The container has also played a part in altering the places in which these goods are made. By dramatically cutting shipping costs, the steel box has rendered geography far less relevant for manufacturers, who go where costs – mainly workers' wages – are low to have jeans or computers made. So for the fish industry, if transport costs represent a fraction of a salmon's retail value, the issue is less about where it is filleted than what the workers are paid to pick out the bones.

Standard shipping containers are not what have permitted the food industry to send dead salmon on their round-trip voyage to China, however. For this task, the container's more sophisticated cousin, the reefer unit, has done the work. To some, the word 'reefer' conjures up the image of a hand-rolled marijuana cigarette (curiously, there may be a shipping link here since 'reefing' the sails means reducing their size by rolling them up). To those in the maritime cargo industry, a reefer is a temperature-controlled container – usually refrigerated. It is a variant on McLean's box that has enabled exotic fruits and vegetables, seafood and shellfish to be shipped across the globe. From the outside, these refrigerated boxes do not look so different from ordinary shipping containers. Yet inside each one is a complex system of coils, wires and electrical fittings, which are managed by a computer that controls everything from the temperature and humidity to ventilation and gas levels, all working to prevent the deterioration of fresh foods over long distances. This giant cooler box has helped reshape global diets, bringing kiwi fruits to rainy England or mangoes to Americans year round.

Barbara Pratt watched a lot of this happen. Today she is a senior executive at Maersk, the giant Danish shipping company, but in the late 1970s when she left college with a background in science, her first job was at Sea-Land, the company founded by McLean and eventually bought by Maersk. At Sea-Land, Pratt's task was to build a laboratory inside a 40-foot container and then to travel around the world with it to find out how refrigerated products fared during their journeys. For Pratt, then twenty-three, the job was a chance to see the world, albeit from a rather unusual perspective. 'I figured, well, I could do this for a couple of years,' she says. 'And then, of course, I got hooked and never left.'

Pratt's lab was equipped with a diesel generator, and hot and cold running water. There were computers, a chromatograph that measured gas concentrations, as well as other pieces of equipment needed for the experiments Pratt and her team would be conducting. A few comforts were introduced to the container, too – a set of bunk beds, a refrigerator and a microwave – for the team were to spend many a night within the metal walls of their moveable laboratory.

Pratt's first voyage was with a shipment of cocoa beans from the Dominican Republic to the US port of Newark via Puerto Rico. She subsequently followed everything from bell peppers and watermelons to pineapples and blueberries all over the world, keeping an eye on their temperature and humidity along the way. On board the vessels, she would sleep in the ship's cabins but when in ports, she and her team, working in shifts, would frequently spend the night inside the container to do round-the-clock monitoring of the computers and other equipment. Pratt spent almost a decade in her mobile laboratory, checking the effects of ventilation, condensation, temperature and different types of refrigerants. 'Essentially my job was to bring technology to what was then an industry built by truckers,' she says. 'It was pretty much the first onslaught of science and technology into the containerised transportation business.'

Before Pratt started making her frosty tour of the world, companies with names like Carrier Transicold (founded by Willis Haviland Carrier, the 'father of air-conditioning') and Thermo King had been busy developing transportable refrigeration units. As with the container itself, the trucking industry had a hand in the development of the reefer. In the summer of 1938, so the story goes, a businessman called Joseph Numero, who then made sound systems for cinemas, was playing a round of golf with his friend Harry Werner, who owned a truck business in Minneapolis. When Werner started complaining that the summer heat was ruining many of his meat deliveries, Numero claimed he could come up with a solution. His golfing partner was highly sceptical. He found it hard to believe Numero would be able to create something small enough to move around that was also durable enough to withstand rough journeys in a truck.

What Werner did not realise was that Numero had a powerful tool at his disposal – the genius of Frederick Jones, a self-taught mechanical engineer who became one of America's most prolific black inventors, his sixty patents including a portable X-ray machine and a racing car. While working for Numero's cinema business, he devised an automatic ticket machine and a movie projector that played back recorded sound. Despite Jones's extraordinary powers of invention, several early attempts floundered. Eventually, with a few bits and pieces he collected from junkyards, Jones put together a rather ungainly unit that was attached to the base of the trailer. The strange contraption worked.

Jones and Numero improved on their 'Model A', with a 'Model B' and a 'Model C' and patented their machines. The pair then went on to found US Thermo Control Company – today's Thermo King, now owned by the Ingersoll Rand industrial group. In 1956, the company's refrigeration unit was installed on an ocean-borne vessel. The new technology effectively turned the shipping container into a mobile fridge that, powered by propane, was self-sufficient. The reefer can be plugged into the electric grid on vessels, at the container yard or into its own diesel-powered generator set. Suddenly, perishable foods could be sent on journeys of thousands of miles through tropical heat without deteriorating.

Another breakthrough came in the late 1970s with the advent of the microprocessor. Until then, reefers had been controlled mechanically, with primitive sensors telling the operator whether or not interior temperature was being maintained. 'When they introduced the microprocessor, all of a sudden you had all these computer capabilities,' explains Pratt. 'You could record all these additional things that were going on inside the reefer and begin to use that information to develop more reliable refrigeration units that did their own maintenance while they were operating.'

Armed with microchips, these boxes are extremely clever. They give individual care to foods that have different travelling requirements – from products with high fat content such as fish, for which colder is better, to stone fruit, avocados and kiwi fruits that must be kept as cool as possible without actually being frozen. Ice cream is a particularly delicate creature. It must travel at a constant, and very low, temperature because every time it melts and refreezes, its water crystals become larger, giving it an unpleasant crunchy texture. Microprocessors allow the reefer to do this, whether the ship is in the tropics or the Arctic. Not only do they keep the temperature constant but they can also control levels of oxygen, carbon dioxide and nitrogen. These gases alter the respiration rate of fresh produce, which in turn slows down the ripening process, extending its shelf life and making it ready to eat at just the right moment – when it is unpacked and loaded on to the shelves of the supermarket.

Microprocessors can even keep an eye on things during the voyage, detecting problems and fixing them – sending alerts to the vessel's bridge or, via satellite communications, to a website through which shippers can make adjustments to the reefer temperature remotely. The computer can be set up to send status reports every half an hour for sensitive cargo. Using the same satellite technology, global positioning systems can track the location of the container to within 10 feet. You wonder if McLean ever imagined that computers in outer space would one day allow shippers to communicate with their cargoes from so great a distance.

When it comes to fish such as our salmon, companies have developed new types of refrigerated containers that can generate the ultra-low temperatures needed. Technology, however, is not the only factor behind the practice of shipping frozen fish to China and back. A crucial element in the economic equation of this incredible journey is the fact that wages in the People's Republic are a fraction of what they are in the west. For the same price as employing a single

Norwegian seafood-processing worker, for example, twenty-five Chinese workers can be hired to pick away at the innards of fish and seafood. Crab shakers – who extract the meat from crab shells with pincers – earn between $100 and $150 a month in Qingdao, about a tenth of what it would cost to pay someone doing the same job in the US.

Because the shipping container has allowed companies to seek out countries with the lowest wage levels, the various parts of a product no longer need to be manufactured in the same place. Each can be made where the processing is cheapest and then assembled somewhere else. Take the auto industry. The farmer in Ohio is probably not thinking about this as he rattles round his land, but parts of the truck he is driving may have been made anywhere from China to Thailand, where many of the parts needed to make the truck are semi-finished before being shipped to the USA in containers. Only the final assembly is carried out locally. In the textile and footwear industries, most manufacturers do not even own their overseas factories but subcontract to local companies.

As a result, some corporations no longer actually make anything – they leave the production to others, and simply brand, design, assemble and market the things they sell. When, for example, car companies cut back on production, the share prices of dozens of other firms start to fall as investors worry about falling demand for their radios, airbags or seats. When fashions change in New York, textile factory managers in Indonesia and Thailand must pay attention – they may need to invest in some additional machinery to produce the new garments and hire some extra workers. We live in a globally connected world where the effect of each new development in one place is felt in another. Such a world could not have been created without the assistance of the shipping container. It is more than a simple, moveable steel box – it is globalisation's porter.

In September 2003, in the Mexican beach resort city of Cancun, Lee Kyung-hae, a Korean farmer, took out a knife and stabbed himself in the heart. Lee's suicide, an act of protest, was not an outcry against a corrupt regime, inhumane treatment of political prisoners, or the suppression of his religious beliefs. He was demonstrating at a meeting of a global body that formulates the rules of commerce among nations, the World Trade Organisation. 'WTO kills farmers!' Lee cried as he plunged the knife into his chest. It was a shocking scene.

Lee's dramatic death was part of a profound questioning of the changing international economic order – a new order that the shipping container helped to bring about by efficiently, speedily and cheaply delivering goods around the globe. The anti-globalisation movement emerged in the late 1990s, when the protests of students and activists began to achieve international prominence. In May 1998, thousands of protesters formed a human chain round the G8 summit venue in Birmingham, England. The following year, at a WTO meeting in Seattle, violence hit the headlines as hundreds were arrested and scenes of vandalism were broadcast across television screens worldwide. The anti-globalisation protesters' grievances varied widely. They ranged from concerns about pollution and dismay at seeing jobs moving overseas to anger at the impact of global trading rules on poor countries and sweatshop conditions in foreign factories. A moustachioed sheep farmer called José Bové vandalised a half-built

McDonald's restaurant in a provincial town in southern France. Everyone from environmentalists to union leaders took part in protests. Even the anarchists showed up.

Clothing and footwear companies were particular targets of the protesters. Ironically, the anti-globalisation activists had at their fingertips a very global tool – the internet. Suddenly, with instant communications available to millions of people, the way multinationals behaved in far-flung corners of the world could no longer be kept from view. Once campaign groups identified human rights violations or poor conditions in the factories, farms and plants used by western companies, word spread rapidly around the globe via the web. Big clothing brands such as Gap and Nike and food companies such as Chiquita, the US corporation best known for its bananas, found themselves at the receiving end of protests, consumer boycotts and critical press reports. The activists' spotlight has not yet fallen on the global fishing industry, perhaps because no well-known brands are associated with it. If consumers become more interested in the origins of the food they eat, the employment conditions of the scalpel-wielding women of Qingdao and others like them may also come under scrutiny.

Even if they do, however, the answers are not easy to find. Companies have suddenly had to become experts in managing factory conditions – no easy matter when you do not actually own the factories where your products are made or your foods are processed. Government efforts to improve the lot of workers in poor countries can backfire. When the Harkin Bill, designed to reduce child labour by prohibiting imports of products made by children, was introduced to the US Congress in 1992, manufacturers in Bangladesh, fearing the loss of a lucrative market, started indiscriminately firing young workers, many of whom fell into worse conditions than they had endured before – some ending up as street vendors or prostitutes.

Today, many recognise that any actions forcing companies to pull out of countries or factories simply put workers' jobs at risk. Big brands have started working with non-profit organisations and auditors to make sure their codes of conduct are being followed by the factories and farms where their jeans are sewn or their vegetables are harvested and packed. From the early 1990s, an unlikely partnership with a New York-based environmental action group called the Rainforest Alliance helped transform Chiquita from a corporation despised by many into a business respected for its efforts to improve the lives of workers. Getting local managers to co-operate is part of the process. One enterprising American executive from a technology company even managed to persuade the bosses of the Chinese electronics plant making her company's products to spend a night in the workers' sleeping quarters – after which she noticed a marked improvement in the state of the dormitories.

As the corporate world has woken up to these issues, global brands are actually emerging as a positive force for workers. After all, the bigger your name, the more you have to lose by being caught participating in abuse. And, as Nike discovered, once your brand is tarnished it is hard to repair the damage. The company has long been seen as a world leader in managing the conditions in its factories around the world, yet many still associate 'Nike' with the word 'sweatshop' – meanwhile, the behaviour of no-name local firms often escapes attention.

While activists continue to scrutinise the actions of multi-national companies, the shipping container is never far away.

But as it is loaded and unloaded at docks around the world the steel box is also creating jobs – often better-paid ones than were available before containerisation entered the scene. The $100 or $150 earned by Qingdao fish-processing workers filleting salmon may not sound like much, but it is higher than what they would earn as farm hands. Globalisation has side effects that are harsh – whether for poorly treated workers on one side of the world or those who have lost their jobs on the other. The number of fish-processing jobs in the US state of Washington has, for example, fallen by more than twenty-five per cent in the past decade. In her investigation of the food industry, Felicity Lawrence found Moroccan horticultural workers on Spain's Costa del Sol living in squalid conditions without access to drinking water or sanitation.

Yet these are problems that should be addressed on land, through retraining programmes, financial safety nets and tough measures to prevent labour abuses. They are not reasons for halting the passage across the oceans of travelling T-shirts or moveable feasts.

Lee Kyung-hae was not protesting against sweatshops with his dramatic suicide. His action was a response to a more complex inequity in the global trading system – protectionism. It is an inequity that is particularly evident in food and agricultural products. Governments use two mechanisms to protect their farmers: subsidies that keep the price of home-grown foods artificially low and duties or tariffs imposed on the foreign produce arriving on their doorstep. What Lee and his fellow farmers were opposing was their government's removal of the high tariffs on foreign imports that had protected Korean rice. Suddenly they were exposed to competition from cheaper imports. Lee became a spokesperson for his fellow farmers, arguing that free trade was destroying their livelihoods – and in some ways it was. The trouble was that Korean rice was expensive. Its price was far higher than that of rice produced elsewhere and it was often being harvested from paddy fields planted by hand, using methods that had not changed for centuries. Without artificial protections, this ancient form of farming could not survive, particularly since other countries such as the USA were still handing out massive subsidies to their farmers.

Korean farmers, who had for years been protected by these mechanisms, were thus casualties of free trade, or rather, unfair trade. Yet free trade was what enabled the rest of their country to transform itself from third-world poverty to first-world prosperity in a matter of decades. When India was one of the most closed markets to international trade, it was also one of the world's poorest. Since reversing its policy of commercial isolationism in the early 1990s, the country's economy has expanded rapidly. Vietnam in the 1980s was a country closed to outsiders. The streets of the capital, Hanoi, were sad and dark. Food markets consisted of dishevelled groups of traders huddled round a few vegetables and meals were eaten on the pavement by the light of small oil lamps. From the mid-1980s, as part of an economic reform policy called *doi moi*, Vietnam opened to international trade and investment. Today Hanoi's streets are lined with cafés and shops, and stylishly dressed women speed around the place on mopeds. No doubt the participation of these countries in world trade will produce casualties such as the Korean farmers, but India and Vietnam are now among the fastest-growing nations in the developing world.

Trade protections exist in wealthy countries, too. In the

USA, the notorious farm bill secured vast and much-criticised subsidies for American farmers, while in Europe farm tariffs remain high and Common Agricultural Policy subsidies are frequently attacked for, among other things, being the equivalent of paying every cow on European soil about $2.50 – a more generous daily wage than that received by more than half the world's population. Without these subsidies, say the governments behind them, many in rural areas would lose their jobs. The reality is, however, that the bulk of US and European subsidies go to large agribusinesses, not small farmers. Moreover those subsidies encourage farmers to over-produce. The surplus foods are then dumped on international markets – and at prices far below the cost of producing them, making it impossible for farmers in Africa and Asia to compete. Subsidies to US rice farmers (which incidentally cost American taxpayers an average of $1 billion a year) bring down world rice prices by up to six per cent. Such policies, say famine relief experts and development agencies, hamper efforts to reduce world poverty.

Of course, agricultural commodities such as rice tend to travel loose, in bulk shipments, rather than packed into shipping containers. However, the container is the means of travel for other items such as dairy products, dried coffee and bananas, and the tariffs put on many processed foods can be even higher than those on raw materials. For all the efficiency of the big steel box, the obstacles to moving our food around are not always purely logistical.

While the sentiments that motivated the anti-globalisation movement of the late 1990s were often confused and violence tarnished the mood of many otherwise peaceful protests, the demonstrations left a lasting impact. By achieving global prominence, the protesters highlighted the plight of the poor and the importance of reforming international trading rules, sparking debates that continue today. The movement reached its most dramatic moment in 2002 when more than 100,000 protesters turned up at a meeting of the G8 industrialised countries in Genoa, Italy. Violent battles between police and demonstrators broke out on the streets, leaving one dead – twenty-three-year-old Carlo Giuliani – and dozens injured. Determined to prevent the protesters from achieving their stated goal of breaching the Red Zone, a high-security area in the city centre where the G8 leaders were meeting, police built barricades around the zone. And what were the objects with which they constructed these ramparts? Hundreds of steel shipping containers – in a curious twist of fate, the most important tool of the global trading system had been transformed into a weapon in its own defence.

Cannon Fodder

Battlefield food fuels packaging technology

Tomato (canned): Fleshy, usually red, fruit (or more accurately, large berry) that has been skinned by dipping in hot water and sealed inside a tin can

Origin: Central and South America

Etymology: From Spanish *tomate*, from Nahuatl *tomatl*

Legends: When the tomato was introduced to France from Italy, the French mistook the name *pomei di moro* – or Apple of the Moors, because it had been brought from Spain by Muslims – for *pomme d'amour* and treated it as an aphrodisiac

*G*erry Darsch is a food specialist who grew up in Massachusetts surrounded by packaging and preservation technology. His uncle, Rocco Fiorini, owned a company in Cambridge that packaged Italian pork products. His father started a business blending salad oils and vinegar, making casings for the meat industry and providing storage for Stella Cheese products. As a child, Darsch was fascinated by everything about the business, from the giant fridges that stored his father's cheeses to the cellulose casings – closed neatly with a metal ring – that were slipped over franks, wieners and Italian salamis. 'He'd clip one end and box them and send them out to the meat packing plants to be stuffed with Italian pork products,' he says. 'I always found it intriguing.'

After finishing high school, Darsch took up his studies, first at the University of Massachusetts Amherst and then at Framingham State College, 20 miles west of Boston. In May 1975, he graduated and a month later, armed with a couple of degrees in food science and plenty of enthusiasm, he landed a job in the food industry. He has remained there ever since. Unlike many in the food business, however, Darsch is not surrounded by chefs or product managers. His colleagues include nutritional biochemists, packaging technologists, mechanical, chemical and electrical engineers, food technologists, behavioural scientists and dieticians. Darsch runs the US Department of Defense's combat feeding programme. He and his team represent the brains behind the ingenious devices that allow US soldiers to eat while on the move.

Darsch's place of work, the Natick Soldier Systems Centre, is hardly what you would expect a military installation to look like. Set beside a lake in a Boston suburb, the 78-acre campus resembles a suburban business park. Nondescript low-rise buildings are dotted about like large shoeboxes on grassy terrain surrounded by woodland and visited by the occasional flock of Canada geese. Architecturally, there is not much to look at, but Natick Labs, as it is known, is a key Pentagon installation. Futuristic gadgets and high-tech equipment emerge from within its laboratories and testing chambers with astonishing frequency.

This is the starting point for the development of equipment and supplies that will end up in the hands of the US Navy, the Marine Corps, the US Air Force and the US Army and, in some cases, the US Coast Guard. To test the new products before they are tried out on the battlefield, two climatic chambers on the campus create tropical and Arctic temperatures and simulate humidity, rainfall and solar load. A machine with a fan 12 feet wide produces sustained winds of up to 40 miles an hour. Inside the testing chambers, volunteers spend their days battling fake winds and artificially generated temperatures. They might be asked to wear patches that collect sweat to measure lost nitrogen. They can be strapped up to heart rate monitors while they walk or run for several hours on a variety of treadmills. Watching videos or playing computer games helps relieve the boredom.

The constant stream of innovations that emerges from this unusual research centre includes ultra-lightweight gloves whose flexible fabric means they pass the 'trigger test', allowing the wearer to continue operating a weapon, and clothes made from selectively permeable membranes allowing sweat to escape while keeping out the toxic agents used in chemical warfare. However, it is the study of food that Natick Labs is best known for. Here, scientists devise the rations that will eventually be eaten by soldiers in the field – sophisticated MREs (Meals Ready to Eat) that heat themselves using a chemical reaction and come with accessories such as tiny bottles of Tabasco sauce and sachets of coffee. They seem to work like magic. Just add water to the granola with blueberries and it turns into a bowl of granola with milk. Put an ounce of water into the pouch that comes with an MRE and it will heat up the food in ten minutes. Rip the tab off the dreary-sounding Unitized Group Ration – Express (the UGR-E), and without the assistance of cooks or equipment it soon transforms into dinner for eighteen, whether that is chicken chow mein or barbecued pork ribs.

The list of requirements for each meal is long – from nutritional value to special menus for cold-weather conditions and packaging that is waterproof, vermin-proof and insect-proof. The biggest preoccupation for developers is coming up with ration packs that are easy to transport and carry. These military moveable feasts must undertake tough journeys through wildly fluctuating temperatures without deteriorating. They may be thrown into the back of trucks and dropped from helicopters without a parachute. Then they must give soldiers individual, nutritional meals that they can, if necessary, consume with one hand while running.

The need for portable rations has become more pressing in recent years. Back in 1991, it was hard to imagine any food was involved in the Gulf War, an information-age conflict that appeared to be dominated by computers, aerial photographs and radar screens and was viewed by citizens as a series of flashing lights on TV screens. To the American public at least, the war looked more like a video game than a bloody battle (though of course the devastation on the ground was very real). That has all changed. The battlefield is becoming 'asymmetric' – that is, without clear front lines. In places such as Iraq and Afghanistan, special operations forces conduct door-to-door urban warfare on foot. Technology is still very much part of the picture, but it is technology deployed in the form of intelligent body armour, advanced communications equipment and an array of high-tech gizmos designed for individual soldiers fighting on the ground.

This means that, more than ever, soldiers need rations that they can carry with them. For Natick, then, the constant drive is to reduce the weight and dimensions of army food. In the process, technicians have come up with some surprising innovations. After all, few people would think of urinating on their food before eating it. That, however, is exactly what soldiers can now do when, in the absence of water, they need to rehydrate their rations. In one of Natick's more bizarre creations, its scientists have developed a pouch with a filter that removes '99.9999' per cent of bacteria and most toxic chemicals from whatever liquid is on hand. Using an ingenious process called 'forward osmosis', the hydration pouch lets only pure water seep through it. This means that, if necessary, soldiers can soak their rations in anything from swamps to industrial waste, cutting the amount of water they have to carry with them by about half.

The rations themselves are constantly getting smaller and lighter. The latest innovation is the Compressed Meal (CM), which takes up one-third less volume and cuts about the same amount of weight from the traditional MRE without losing any of the accessories or calories. Another is the First Strike Ration, a large cylindrical sandwich rather like Nestlé's Hot Pockets, the stuffed hand-held convenience products found in the frozen food section. While the MRE is a heavy brick-sized pack with several components and requiring some preparation and a spoon, the FSR (the military loves its acronyms) is a single hand-held package of high-energy food that feeds a soldier for seventy-two hours yet weighs about as much as one and a half MREs and cuts back on wasteful 'field stripping' – a habit whereby soldiers lighten their load by throwing away the bits of the ration pack they do not want.

Natick's biggest breakthrough came in the 1970s. The idea was to replace heavy, inconvenient cans with a single package that could easily be stashed in a soldier's pack and whose contents would remain safe to eat without being refrigerated. This meant heating the food, since the micro-organisms that destroy organic material can only survive at certain temperatures. Tests had shown that a flat pouch with greater surface area than a cylindrical can would require less heat in the sterilisation process, leaving the food with better taste and texture. The development of the portable pouch was a complex business, however. Four layers of polymer were to be used, as well as a membrane of aluminium foil, but an adhesive had to be found that would hold the layers together without leaching glue into the food. The pouch then had to be flexible and lightweight and yet able to withstand the fierce heat used in the food sterilisation process and the rough handling of the military logistics system. It had to be able to be manufactured on a massive scale. Moreover, it had to meet US Food and Drug Administration approval. The scientists and engineers got there in the end. Going by a less than glamorous name – the 'retort pouch' – the new ration pack was by the 1980s being introduced into the military logistics system. In sludge brown, with small white labels describing the contents, these dull-looking flexible packages now contain virtually all meals eaten by US soldiers. Sealed at the edges and opened by tearing off the top, they are warmed by an exothermic, or heat-releasing, chemical reaction that is activated by adding a little water.

Even with the assistance of the retort pouch and a chemical heater, the burden of feeding US forces remains a heavy one. The Defense Supply Center Philadelphia is the starting point for all military supplies. From this austere red-brick building, which in the nineteenth century was a warehouse for ammunition and other supplies, food is ordered from three US-based companies. While many of the rations can be kept in warehouses around the world or on marine pre-positioning squadrons (giant cargo vessels that carry tens of thousands of meals, many of them in shipping containers), much of the food needed for armed forces and support staff has to be hauled into the areas of conflict – and in large quantities. During the Gulf War in 1990, the US military was shifting supplies of food out to its troops that would have kept the entire city of Atlanta well fed.

Of course, all this hassle could be prevented by doing away with feeding troops altogether. This may sound like an odd concept but it is something that military scientists are getting excited about. One idea is that, when fighting gets in the way of dinner, soldiers could get everything they need through a transdermal nutrient delivery system (a TDNDS, of course). Rather like a nicotine patch, this would slowly release a cock-tail of nutrients into the body, giving soldiers enough energy to get through a battle. Using sensors to determine the soldier's requirements, a microchip embedded in the patch would activate a system that would transmit the necessary nutrients – possibly using electrical impulses to open the pores or, through microdialysis, feeding nutrients straight into the blood vessels. Natick reckons the system might be ready for use around 2025.

Even more futuristic are proposals for battlefield nourishment being considered by the Defense Advanced Research Projects Agency (Darpa), the Pentagon's research centre. Darpa calls it 'metabolic dominance', and if that sounds rather ominous, perhaps it is. The idea is that, using biochemistry, soldiers could be enabled to remain mentally alert and physically able for up to five days without food. Various avenues are being explored in what is called the Peak Soldier Performance Program. One theory is that keeping the body at a certain temperature could enhance physical performance. Even further out on the experimental edge of science is the idea that the mitochondria present in human cells could be made to perform better. Mitochondria convert nutrients into energy, and research has shown that endurance athletes possess greater numbers of mitochondria and can process carbohydrates and lipids, or fats, more effectively than someone who sits around all day doing nothing. Scientists at Darpa wonder whether it would be possible to increase the number of mitochondria per cell and to change what they metabolise so that they could tap into the abundant, unused fat stores in the human body.

Some suggest that there could be more sinister uses for such biochemically driven feeding techniques. In *Perfect Killer*, a fact-based military conspiracy thriller with an extensive bibliography, novelist Lewis Perdue focuses his action on the quest of Bradford Stone, a renowned neurosurgeon, to find out who is behind a secret plan to dope US soldiers with Xantaeus, a drug that 'robs a soldier of free will without their knowledge, overrides their sense of compassion and neutralises the fear of injury'. The new drug, Perdue writes, would be administered through a trans-dermal patch. As Perdue explains on his website, *Perfect Killer* was inspired by *No More Heroes*, a non-fiction book by Richard Gabriel, a professor at the US War College who examines the centuries-old mission to find a drug without side-effects that eradicates fear and the reluctance of humans to kill others. The advent of the chemical soldier, argues Gabriel, will change 'not only the nature and intensity of warfare, but the psychological nature of man himself'.

Still, until scientists work out how to manipulate mitochondrial function, soldiers need real food – and that means coming up with breakfast, lunch and dinner menus that are nutritious, appetising and moveable. Once the rations have been crammed with the right kinds of nutrients and calories to ensure soldiers can operate at their peak, Natick scientists turn to the more creative part of their job – making it taste good. This is no small matter. Military strategists know that what does not taste good is thrown away, reducing the soldiers' caloric intake, which can dramatically affect performance.

For many years, the unappetising nature of army rations did little to help. During the American Civil War, soldiers on the march had to chew on 'sowbelly', a piece of salt pork, and 'hardtack', a thick white-flour cracker so hard to bite into that it was known as a 'teeth-duller', and which was often softened by soaking it in coffee or frying it in grease to create a 'hellfire stew'. As late as the 1990s, soldiers were giving their rations derogatory names. During the Gulf War, the culinary reputation of MREs was so low that they were known as 'Meals Rejected by Everyone', 'Meals Resembling Edibles' or 'Meals Rarely Edible'. Even the latest high-tech rations come under fire from those that have to eat them day in, day out

('Gag in Bag' is one of the less flattering nicknames), and doctoring MREs or bartering the best bits remains a common practice.

Still, compared to earlier rations, today's MRE isn't so bad. A change in strategy from what Gerry Darsch calls the 'father knows best' approach to one where 'customer' surveys identify soldiers' preferences has led to a more varied military menu and a growing selection of ethnic dishes such as teriyaki chicken, shrimp jambalaya and pork tamales. Some of the menus sound positively appealing, such as the oriental chicken in Thai sauce with yellow and wild rice pilaf followed by raisin nut mix, peanut bar and cracker and French vanilla cappuccino. And, as it turns out, the meals are not that bad. The Mexican corn and the meatballs with marinara sauce are surprisingly tasty and the bacon is astonishingly crispy considering it has come straight out of a plastic packet and may be several months old. Haute cuisine? No, but it is probably pretty good when you are sitting caked in dust in the burning heat of a desert somewhere in the Middle East.

Soldiers can even eat junk food – the MRE now includes items such as M&Ms, Oreo cookies and Jolly Rancher Candy in their original packaging. Careful calculations lie behind the introduction of these foods. While most meals still come in sludge brown camouflage packets (plus white for the cold weather versions), field studies have shown that 'a taste of home' acts as a significant morale booster. Catering to captives has led to menu changes. The vegetarian MARC (Meal, Alternative Regionally Customized) was originally designed for detainees at Guantanamo Bay. Different faiths are even accounted for with special meals for soldiers who maintain a strict religious diet – although, of course, the Jewish kosher meal varieties are identical to the Islamic halal options. Purchased separately from different companies with certified preparation methods that adhere to each religion's requirements, they are separated by just a single digit in the order number: 8970-01-E10-0001 for the kosher ration and 8970-01-E10-0002 for the halal version.

'We try to please everyone all the time,' says Darsch. 'We have 1.2 million war fighters that are our customers, and war fighters carry automatic weapons and pistols – you don't want to antagonise a customer base like that.' Darsch grins as he delivers one of his favourite lines. But he is deadly serious about the combat feeding mission. Darsch argues that while much talk about modern warfare revolves around gadgets such as global positioning systems, chemical agent monitors and field computers, food remains a crucial battlefield tool. 'The US has the most lethal weapons platform in the global arsenal,' he says. 'But those weapons are only as good as the individual war fighter that operates them – our job is to fuel that war fighter.'

Army strategists have long been aware of the vital link between mobile food and military prowess. Throughout history, legends tell of the heroic clashing of swords and the charging of steeds, but behind them were great carts of food bringing up the rear. Feeding armies was a gargantuan task. When Charlemagne, the ninth-century French emperor, summoned troops from across the empire, the monastery of Saint Wandrille was among those called upon to assist. Its abbot had to provide about 850 men, plus a three-month supply of food for each one. This meant marching with 75 tonnes of grain on more than 150 carts, each drawn by two oxen. Along with transport for equipment required for battle, this created a baggage train more than a mile long.

As well as feeding men, there were also the horses to worry about. This is easy enough to manage when there is plenty of grass at hand. However, when America launched an attack on the British in Canada in 1775, one reason behind the initiative's failure, argues historian John Shy, was the fact that General Philip Schuyler had to delay the advance until late summer rains had nourished the grass needed to feed his animals. Because the campaign was launched too near to the winter, the harsh Canadian cold forced the Americans to retreat. During the Boer War in South Africa, the English, finding no suitable horse fodder locally, imported it from Latin America. As a result, the war changed not only the course of South Africa's history, it also altered the ecosystem, since the grass seeds took root and what is now known as Khaki Weed, because of its wartime origins, runs rampant, much to the chagrin of local farmers.

As well as looking after their own army's needs, military planners have long recognised the decisive advantage often brought about by seizing the enemy's rations. In *One Hundred Unorthodox Strategies,* a Chinese military handbook compiled in the fifteenth century during the Ming dynasty, we find this advice: 'If you occupy the enemy's storehouses and granaries and seize his accumulated resources in order to continuously provision your army, you will be victorious.' What is more, until the advent of the MRE, it was not only the food itself that was of strategic importance. After all, what use is grain unless it can be turned into bread? French strategists were quick to recognise the importance of cooking equipment. In order to hamper the advance of attacking Spanish armies, a seventeenth-century directive instructs generals to 'send out before them seven or eight companies of cavalry in a number of places with workers to break all the ovens and mills in an area stretching from their own fronts to as close as possible to the enemy'.

These days, while cooking still goes on at military bases and camps, the rations needed to sustain mobile soldiers are pre-cooked and require, at most, a little heating. These meals have never been inside a refrigerator and can last for years without spoiling. All US combat rations have a minimum shelf life of three years at 27° Celsius or six months at 38°. The military is investigating all kinds of technologies to improve the quality of these long-life meals. Ultra-high-pressure processing, for instance, works by exerting extreme pressure on food. A sealed pouch of food is heated to about 80° Celsius and placed inside a very strong steel container. Water is then pumped into the container and up to 130,000 pounds per square inch of water pressure is exerted on the food. Because the pressure is applied uniformly, the food keeps its shape (if you squeeze a grape between two fingers it can be broken, but if the grape is squeezed from all sides simultaneously it will not be damaged). Because heat is used only for a few minutes and the entire process takes only twenty minutes (as opposed to about an hour with traditional sterilisation), the food retains its fresh-ness. The technique, which deactivates any bacteria, could help produce rations that look, feel and taste better. Great hopes are pinned on the process as a solution to producing egg dishes, which are notorious for ending up rubbery and tasteless.

For now, however, most of today's ration meals are sub-jected to what has become the standard sterilisation technique whereby they are put in a retort pouch and, for ninety minutes, subjected to a temperature of 120° Celsius. The MRE may be produced using high-tech equipment but the process used to give the food its three-year shelf life is actually based on a method of preservation that was developed in early-nineteenth-century France.

Nicholas Appert was born in 1750. Now revered by the food industry as the 'father of canning', he found that heating food in vacuum-sealed containers destroyed the micro-organisms that cause it to go bad. Appert, a chef and confectioner, experimented with champagne bottles (his choice of vessel is hardly surprising, since he was living in the small town of Chalons-sur-Marne – today's Chalons-sur-Champagne – on the fringe of the Champagne region). He filled his bottles with fruit, vegetables and meat, sealed them with corks and submerged them in baths of hot water. The results of his work gave the world a means of preserving food that has been used ever since.

In his book (its snappy title was *The Art of Preserving All Kinds of Animal and Vegetable Substances for Several Years*), Appert described the process:

> First, enclose the substances you wish to preserve in bottles or jars; second, close the openings of your vessels with the greatest care, for success depends principally on the seal; third, submit the substances, thus enclosed, to the action of boiling water in a bain-marie for a period of longer or shorter duration, depending on their nature and the manner I shall indicate for each kind of foodstuff; fourth, remove the bottles from the bain-marie at the appropriate time.

It was a set of instructions that was to have profound implications for the food industry. By discovering that food remained fresh when sealed in an airtight container and immersed in boiling water for a few hours, Appert had identified a principle that is at the heart of modern food preservation techniques.

Unlike the scientists at Natick, however, Appert tackled the challenge of food preservation from the kitchen, not the lab. After a stint working for nobles such as Duke Christian IV of Deux-Ponts, Appert moved to Paris where, at the age of thirty-one, he established a successful business as a confectioner and began examining the effects of sugar on food preservation. In 1794, he sold his shop and moved to Ivry-sur-Seine, where he eventually became mayor. However, it was his work with food preservation that took up much of his time. Under the patronage of Alexandre Balthazar Grimod de la Reynière, a financier whose books on gastronomy had introduced the French middle classes to the delights of cuisine, Appert set up a factory in the village of Massy, hired fifty workers and, with the assistance of large copper vats, started rolling out his bottled food on an industrial scale.

While he continued to experiment and refine his methods, Appert was a chef, businessman and culinary connoisseur rather than a scientist, and he was driven by frustration at the poor taste and texture of foods preserved through methods such as salting, drying or smoking – techniques that had been used since Roman times. Although he became a highly skilled food technician working on sterilisation techniques that foreshadowed Louis Pasteur's discovery of pasteurisation, Appert probably did not use any scientific reports to inform his food preservation trials. Nor did he realise that what he was doing as he stood over his boiling pots and corked bottles was developing and refining ideas that had been explored in previous centuries.

Eighteenth-century fruit bottling was one example. In *The Art of Cookery Made Plain and Easy*, published in 1747, Hannah Glasse, an English housewife living in Bloomsbury, described the preservation of gooseberries:

> Pick your large green Gooseberries on a dry Day, have ready your Bottles clean and dry, fill the Bottles and cork them, set them in a kettle of Water up to their Neck, let the Water boil very softly till you find the Gooseberries are coddled, take them out, and put in the rest of the Bottles till all is done; then have ready some Rosin melted in a Pipkin, dip the Necks of the Bottles in, and that will keep all Air from coming in at the Cork, keep them in a cool, dry Place, where no Damp is, and they will bake as red as a Cherry.

Scientists, too, had picked up on the connection between heat and sterilisation. In the mid-seventeenth century, Robert Boyle, the British chemist and intellectual, conducted an early exercise in vacuum packing when he preserved a number of different foods 'in vacuo' using an air pump. In 1679, the French physicist Denis Papin, who spent a number of years in London working with Boyle, developed a sort of pressure cooker called a steam digester. Others busily debating food preservation from the microbiologist's perspective included Italian priest Lazzaro Spallanzani and his adversary Englishman John Needham. Though disagreeing on the theory of spontaneous generation (that life forms arose spontaneously from non-living matter), both suggested that boiling food and excluding the air would prevent microbes from growing.

It is unlikely that Appert followed such debates. He preferred to concentrate on his cauldrons and bottles, stuffing anything he could lay his hands on into jars. 'Without apparently being aware of it,' writes historian Sue Shephard, 'Appert had brought together the different discoveries of the past.' Appert began to receive recognition for his work. The *Courrier de l'Europe* recognised the potential of his techniques. In February 1809 the paper told its readers: 'M. Appert has found a way to fix the seasons; at his establishment, spring, summer and autumn live in bottles, like those delicate plants protected by the gardener under glass domes against the intemperate seasons.'

For Napoleon such romantic sentiments were beside the point. He was determined not to let food get in the way of his ambition. 'An army marches on its stomach,' he famously declared. Mobile rations, Napoleon knew, would allow his troops to travel faster and further, thus extending his domination of Europe. In his pursuit of conquest, Napoleon had already revolutionised battlefield logistics. By breaking the 'umbilical cord' of supply-bound armies, forcing his troops to forage for themselves, he made his army far more nimble. He also brought order to foraging techniques that had until then been little more than chaotic pillage. By broadening his front, he created a wide area from which the troops could send organised forage parties. 'You must order people to provide oats, and confiscate the oats from those who do not,' Napoleon wrote to Prince Eugène de Beauharnais, Viceroy of Italy, on 22 September 1805. 'Do not worry about these measures displeasing the country. People whine, but they do not mean what they say. They know full well that in circumstances such as these one cannot do otherwise . . . Do not be offended by anything. These moments are moments of suffering.' Napoleon's method of securing supplies might have caused consternation, but freedom from supply lines meant his army could move rapidly, giving him a significant strategic advantage.

In 1795, the government-funded Société d'Encouragement pour l'Industrie Nationale offered 12,000 francs to anyone who could come up with a practical method of preserving the foods of soldiers and sailors. By that time, the success of Appert's sealed jars of food made him well qualified to win the money. From 1804, he had been sending his preserved foods to be tested by the French navy, which sent back positive reports. In 1810, after giving a demonstration to a government commission that then tasted the results a month later, Appert was awarded the prize.

The money was given, however, on condition he make his preservation techniques available to all. In 1810, he published his book *The Art of Preserving*. Poor Appert. He had not patented his invention. Just three months after the book's publication, across the Channel in England, London broker Peter Durand secured the patent for a sterilisation technique suspiciously similar to Appert's. Preserved food – in tin cans rather than glass bottles – was soon being produced by a factory in Dartford owned by Brian Donkin and John Hall. The sad end to Appert's story is that, despite being declared a benefactor of humanity in 1822 and given a grant to finance his experiments, the money covered little more than the new equipment he needed to purchase. Eventually his wife left him and, abandoned by everyone, he died in 1841 and was buried in a pauper's grave.

Walking around a supermarket, today's shoppers are probably not spending much time thinking about soldiers' rations. However, the military's influence is very much in evidence in the aisles. A growing number of high-tech convenience foods have moved from the battlefield to the marketplace. Increasingly, much of what we buy has been sterilised using the high-pressure processing technology being developed by the US military, particularly cooked meats. Lobster companies are now able to sell meat within the shell and use high-pressure processing to break down cell membranes, making it easy to remove. The same principle can be applied to oysters, since high-pressure processing breaks down the membranes that seal the shells shut, making light work for the oyster shuckers.

Earlier technologies are still in evidence, too. On one aisle, those jars of raspberry jam and thick cut marmalade have been hermetically sealed using the sterilisation principle developed by Appert that proved of such great interest to Napoleon and his British military counterparts. On another aisle, tins of peas and carrots or soup owe their existence both to Appert's work and that of his British counterparts, Durand, Donkin and Hall. Cylindrical metal cans proved much more portable than Appert's jars and bottles and, again, the military took note. By 1813, Donkin and Hall were producing 'tinplate canisters' of food for the British army and navy. One story has it that some were even sent to St Helena, the British island off the west coast of Africa where the exiled Napoleon would end his days. In America, the Civil War gave the tin can a significant boost. Before the war, a new technique was developed that made the cans cheaper and easier to produce and so, when the conflict erupted in 1861, tin cans were available to meet the demand for transportable preserved food. In the period immediately after the war, production of tin cans shot up to thirty million a year from just five million before the outbreak of fighting.

Tin cans now contain everything from soup to soda. Americans use more than 200 million cans a day. Happily, cans made of aluminium (which never wears out) are eminently recyclable – and they can return to the supermarket as a new can in about sixty days, which means a shopper could in theory buy the same piece of aluminium six times a year. There are few foods that cannot be canned, and in some cases they can be better for you. Tinned tomatoes contain more lycopene, a powerful antioxidant, than their fresh counterparts.

But if war helped give birth to the sealed glass jar and the tin can, it may also be responsible for their demise. The retort packaging developed by Natick is being taken up by a growing number of food companies and more of what appears on supermarket shelves now comes in retort pouches, including things that might have once gone in tins or jars. Everything from luncheon meat, tuna and smoked salmon to cooked rice and prepared meals is finding its way into these new flexible packages. The pouches are often designed like sleeves with a flat base that allows them to stand up. Some even have resealable zippers. Like their military counterparts, the pouches consist of an inner layer that is in contact with the food, an outer layer of nylon or polyester – which can be printed on – and a middle layer of aluminium foil that is the barrier to oxygen and water. Crucially, the aluminium can be replaced with new types of barrier films so that the contents can be seen and (perhaps more importantly as people turn to convenience foods) be put in a microwave. As well as being flexible, pouches are easier to open and, most important for the food industry, they take up less 'real estate' in the supermarket.

As a result, tomatoes may be the first foods to disappear from tins. Tetra Pak, a Swedish packaging company that has made the Rausing family among the world's richest, has already started supplying Sainsbury's, the UK supermarket chain, with tomatoes in its 'Tetra Recart' cartons, rectangular cardboard packs that use similar technology to retort pouches. Once heated and sealed, the foods last between six and eighteen months. Among other things, the packaging will, points out the company, save cuts from cans on British fingers, which according to the Royal Society for the Prevention of Accidents amount to more than 2,000 a year.

Few of us wonder how it is that the chopped tomatoes we are putting in our shopping basket – whether in tins or cartons – remain unspoiled even though they have not been in a refrigerator. Most of us remain unaware that the foods we are buying come in containers called retort pouches or that they were developed by the US military. However, one item from the soldier's menu has made it on to supermarket shelves with its military origins clearly visible. The Hooah! bar, an energy bar originally designed to enhance the performance of soldiers on the battlefield, is now being sold to the American public in stores such as Wal-Mart and 7-Eleven. Encased in an eye-catching silver wrapper, it tastes rather like a granola bar and comes in flavours such as chocolate crisp, peanut butter and apple cinnamon. It contains the same nutrients as the original – and is every bit as portable. Some say 'hooah', a widely used army expression, comes from 'HUA' or 'Heard, Understood, Acknowledged'. Others like to think of it as a battle cry. Whichever is true, the Hooah! bar wears its military colours prominently on its sleeve. With red, white and blue emblazoned across the packet, the bar certainly looks patriotic. There is a 'support the troops' ribbon on one end and in large letters on the other are the unmistakable words: 'Created by the U.S. military'.

4

Plane Fare

The Berlin Airlift secures a Cold War victory

Chewing gum: Flavoured preparation, traditionally made of chicle, a milky juice that when dried constitutes an elastic gum

Origin: Prehistoric men and women may have chewed lumps of tree resin

Etymology: French *gomme*, from Latin *gummi*

Legends: On a flight, Philip Wrigley, son of William Wrigley, founder of the chewing gum company, was asked why he continued to advertise a product that was already so famous. For the same reason, he replied, that pilots keep the engines running when the plane is already 29,000 feet up

*I*T IS JANUARY 1949 and, every lunchtime in their one-room apartment in West Berlin, forty-one-year-old Kurt Werner, his wife Charlotte and their three children, Wolfgang, Jürgen and Ines, sit around the table in front of a simple meal. On most days it consists of a stew of potatoes, noodles and – if they are lucky – some meat. In the evening, eating by the light of an oil lamp, dinner might be a bowl of flour soup accompanied by bread (perhaps with a little butter) and followed with a cup of coffee. Breakfast in the Werner household is an even plainer affair – a few slices of bread smeared with butter or fat. However, the Werners consider themselves fortunate to be eating at all. West Berlin, surrounded by Soviet-controlled territory, has been virtually cut off from the outside world for more than five months. They rely entirely on food flown in by British and American pilots. Round the clock, the planes land on the city's airfields, Tempelhof, Tegel and Gatow, loaded with sacks of flour, tins of coffee, cans of corned beef and packets of dried eggs – just enough of the essentials to keep Berliners alive through the harsh winter. It is hardly a feast, but it is certainly a moveable one.

The Werners' dinners – described by a Berlin-based *New York Times* correspondent – probably differed little from those of most West Berliners at the time. However, Charlotte Werner's comments provide a poignant insight into what it feels like for a family to rely on the sky for sustenance. Jürgen, the younger of her two boys, spends the day leaning out of the apartment window, she says, gazing at the streams of aircraft swooping across the city. 'At first, the noise of the planes kept us awake at night,' she explains. 'But now we sleep through it all. It is only when it is quiet that we wake up – afraid the

Luftbrücke [air bridge] has stopped.' For Charlotte Werner, silence is not golden. The constant drone of aircraft is a reassuring sound, signalling the arrival of new supplies. 'Every bite of food we get is flown in by those planes,' she says. Silence might herald the end of the airlift and loss of the vital deliveries arriving in West Berlin. Then, unless citizens were prepared to starve, they would have to surrender to the Russians.

This was the Berlin Airlift. It marked a turning point in post-war history. The operation lasted fifteen months and, at the end of it, the western allies had forced Soviet Russia to back down, kept West Berlin free of Russian control and so hampered the spread of communism through Europe. All this was achieved without guns or bombs. The battle was won with food – airborne food. This military mission had military-style titles. The Americans called it Operation Vittles, while the British name was Operation Plainfare (a misspelling of the word 'plane' in an official document that stuck). But in 1948 and 1949 it was citizens, not soldiers, who were being fed.

Nothing quite like it had ever been seen before – or has been seen since. A community of more than two million was provided with daily nourishment using aircraft deliveries alone. Tens of thousands of people participated in the opera-tion, from the pilots and the cargo handlers to the mechanics in Liverpool, who checked the aircraft every 200 flying hours, and the training officers stationed in Montana who taught precision flying techniques. When Stalin finally ended his blockade in May 1949, not a shot had been fired, but more than 500,000 tonnes of food had been flown into West Berlin (the airlift continued for another four months after the end of the blockade). Along with coal and other supplies, the total tonnage shifted by plane was more than 2.3 million. By pulling off one of the toughest logistical challenges in history, the western allies had not only kept a city alive, they had also secured the freedom of western Europe. It was their first victory in the Cold War.

At the end of the Second World War, Germany was carved up into four occupation zones – British, French, American and Russian. Sitting in the middle of Soviet territory, Berlin was itself divided. Its eastern sector was controlled by the Russians, with West Berlin in the hands of western allies: the British, the French and the Americans. Britain was under the leadership of Clement Attlee, the prime minister, with the single-minded foreign minister, Ernest Bevin, at his side. In Washington, President Harry Truman was at the helm. Western leaders had been watching with alarm as Soviet-style communism marched across Europe at a vigorous pace. The populations of countries such as Poland, Czechoslovakia, Hungary, Yugoslavia, Bulgaria and Romania were already living under Russian rule. It was Germany, however, that was the key to Stalin's expansion plans. If Berlin were to fall to Stalin, western Germany would follow – and, as Lenin once put it, 'whoever has Germany has Europe'.

By 1948 tensions were high. The western powers had declared their intention of keeping their half of Germany within the western European fold. Stalin believed that this could create an obstacle to his plan to sweep across Europe unhindered, so when the allies demonstrated their resolve by announcing the introduction to West Germany of a single currency that would be outside Soviet control, it rattled the Soviet leader. On 24 June 1948, just days after the new currency was introduced, word came that the Russians, who had been gradually limiting access to the western half of the city (citing 'technical difficulties'), had blocked all land and water routes. West Berlin was now cut off from the western world.

Stalin was implementing a well-tested strategy of battlefield brutality: the siege. Cut people off from the food supply, and they soon emerge, hollow-eyed and haggard from the ravages of hunger, to give in to your demands. The allies, Stalin reckoned, would hardly be likely to watch Berliners starve, so would soon surrender their part of the city – and possession of the city was the linchpin in Stalin's plan for a Europe dominated by the Soviet Union.

For those watching events from Whitehall and Washington, news of Stalin's blockade was ominous. They, too, understood the pivotal role of Berlin in Europe's geopolitical future. General Lucius Clay, commander of the US occupation in post-war Germany, was determined not to let the Russians gain control of the city. 'If we mean that we are to hold Europe against communism, we must not budge,' he told General Omar Bradley, the US army chief of staff, in April 1948. West Berlin had to be kept from the Soviet grasp at all costs – but how? The only means of access available to the western allies were three air corridors into Berlin. These routes were secure, since the Russians and the western powers had agreed to guarantee access to Berlin by air at the 1945 Potsdam Conference, where post-war peace settlements were arranged. Any attempt by the Russians to interfere with British or American planes in this strip of airspace would have consti-tuted an act of war, and all-out conflict was something few believed Stalin was likely to countenance since the Soviet Union still feared America's military and economic might.

Running from Hamburg, north-west of the city, Hannover, due west, and Frankfurt in the south-west, these air corridors converged, forming a giant arrowhead pointing to Berlin – and to a possible solution. The air corridors were the only option if the allies wanted to keep West Berlin fed. Prospects for an airlift were discussed. Already, the 'Little Lift' of April 1948 had for ten days taken supplies into the city as Stalin had gradually reduced transport lines into West Berlin. Still, there was great scepticism about the viability of a full-blown airlift. An Associated Press dispatch of 25 June reported that allied experts had said flying sufficient food into the city 'would prove unworkable in the long run'.

The experts' caution was understandable. After all, the amount of goods it was estimated would need to be flown in to keep Berlin alive was impossibly large. The total weight of the daily supply of food would need to be more than 1,500 tonnes, and that was before fuel and other essential supplies were added. Yet the maximum payload of a C-47 was just 3 tonnes. Moreover, the air corridors to Berlin were each only 20 miles wide, so to transport sufficient loads would require so many planes to fly at the same time that the potential for accidents would be enormous. No wonder the idea of hauling all this into the city by air seemed to some sheer madness. On the other hand, things were looking bad in Berlin. Food supplies would soon run out. On 26 June 1948, with no other option in sight, Clay ordered the US Air Force to start flying food and other supplies into West Berlin. The airlift got under way.

Frank Howley, commandant of the US Berlin sector at the time, recorded the tentative birth of what he called 'the boldest and most spectacular supply operation' in aviation history. In his 1950 memoir, *Berlin Command*, he recounted how Clay called a staff conference to discuss efforts to break the blockade. Clay told Howley a handful of planes would be available and asked what he wanted brought in first. Howley chose flour, a food of high nutritional value that was easily handled. The following Monday, 200 tonnes of flour flew to Berlin from the American base at Frankfurt-Main, more than 200 miles away. Howley described the arrival of the first flights:

> They wobbled into Tempelhof, coming down clumsily through the bomb-shattered buildings around the field, a sight that would have made a spick-and-span air parade officer die of apoplexy, but they were the most beautiful things I had ever seen. As the planes touched down, and bags of flour began to spill out of their bellies, I realized that this was the beginning of something wonderful – a way to crack the blockade. I went back to my office almost breathless with elation, like a man who has made a great discovery and cannot hide his joy.

Despite Howley's elation, a daunting task lay ahead. Careful weight calculations had to be made as well as judgements on the minimum amount needed to keep West Berliners alive. There was no space for non-essentials. This was what planners initially settled on as Berlin's daily ration:

> 646 tonnes of flour and wheat
> 125 tonnes of cereal
> 64 tonnes of fat
> 109 tonnes of meat and fish
> 180 tonnes of dehydrated potatoes
> 85 tonnes of sugar

11 tonnes of coffee

19 tonnes of powdered milk

5 tonnes of whole milk for children

3 tonnes of fresh yeast for baking

144 tonnes of dehydrated vegetables

38 tonnes of salt

10 tonnes of cheese

Yet debates continued throughout the lift as to how best to achieve the balance between weight, bulk and nutritional value. While 'DHP' (dehydrated potato) became a staple, planners questioned whether to take up precious cargo space with dehydrated vegetables (favoured by the French and Germans) or the fertilisers that would allow fresh produce to be grown in Berlin's gardens, something the British, whose wartime slogans had included 'Digging for Victory', thought was a good plan. The idea of flying in bread was abandoned as one-third of the average loaf consists of water. It was better therefore to fly in flour that could be turned into bread at its destination. On the other hand, manufacturing ersatz coffee (made mostly of chicory) locally would use up valuable fuel (which also had to be flown in). Coffee, it was decided, would be among the rations. Caffeine must have been a highlight for its recipients. With dehydrated foods and tinned meat ultimately favoured over fresh produce, it was an unappetising shopping list.

The British title for the airlift may have been the result of a typo, but it was more than appropriate. This was plain fare indeed. Quantities were limited, too, as the food was rationed, allowing each West Berlin resident to buy only what was judged sufficient to sustain life. Even when, well into the

plane fare

operation, Major General E.O. Herbert, commandant of the British sector in Berlin, could announce in October 1948 an increase in the daily rations for West Berliners from 1,800 calories to 2,000, this hardly constituted a generous diet (today, reducing your intake to 1,800 calories a day is deemed an efficient way to lose weight). 'In school, we children received a watery soup,' recalled Helgard Seifert in Wolfgang Samuel's collection of memories, *The War of Our Childhood*. 'The teacher made us feel guilty about eating the soup. She would say, "Did you eat well? I didn't have anything."' It was a far cry from the Thai oriental chicken or Mexican-style meat balls that make up today's US military rations. Moreover, it was not only humans that were to experience deprivations. Animals, too, suffered. On 11 August 1948, *The Times* of London told British readers: 'The food difficulty in the Berlin zoo has become acute, and it is now possible to feed the small colony of lions and bears only on alternate days . . . with all sources of supply now cut – including access to the zoo's grass meadows in the Russian zone – it is feared that several of the animals will have to be killed.'

Despite the hardships, West Berliners remained determined not to give in to Stalin. Their hatred of the Russians was intense. In 1945, immediately after the fall of Berlin, Soviet forces had occupied the city for eight weeks. It was a harrowing time. Russian soldiers looted houses and shops and raped hundreds of women. So when Stalin made a bid in August 1948 to encourage residents of the western sector of Berlin to move east, few were persuaded. Only about 19,000 of West Berlin's more than two million residents took up his offer of supposedly abundant food supplies in return for registering in the Soviet sector. A month later, when Berlin's mayor elect,

Ernst Reuter, and other political leaders called a rally to demonstrate unity against Soviet forces, more than 300,000 people turned out. Most West Berliners had clearly decided that signing up to the Soviet way of life was not a price worth paying for a meal or two. And if they had to be blockaded, they would joke, it was far better to be blockaded by the Soviets and fed by the Americans than blockaded by the Americans and fed by the Soviets.

The Berliners did receive some relief from the monotony of their rations. Any spare space between the sacks of flour and boxes of dried egg was taken up by sets of sturdy brown cardboard boxes. From the outside, these were not much to look at, but the impact these boxes had on their recipients was profound. They carried provisions that, for those living in the rubble-strewn ruins of post-war Berlin, constituted unimaginable luxury – items such as corned beef, bacon, margarine, lard, apricot preserves, jams, raisins, chocolate, sugar, egg powder, milk powder and coffee. They were called Care packages.

The packages, which were sent all over Europe after the war, reached people who were often near starvation. In Emmy Werner's *Through the Eyes of Innocents*, people who were children at the time describe the effect of receiving one of these packages. In 1946, Trudy was living as a refugee in Bavaria with her mother and four siblings (her father was in a detention camp). Most of the time, they were forced to forage for berries and mushrooms. It was a bitter winter and the snow lay thick on the ground. When it came to wild food, there was little to be found. So when, just before Christmas, a Care package arrived for the family, there was great excitement. 'Our room became bright and warm with the incredible treasures spilled all over our table,' Trudy recalled. 'Sugar, (peanut) butter, flour, canned meats and fruit cocktail, coffee and chocolate – REAL COFFEE AND REAL CHOCOLATE!!!'

The package was a 'gift from an unknown stranger in America', a clever idea devised by a coalition of charities called the Co-operative for American Remittances to Europe, or Care (now one of the world's largest humanitarian relief organisations). It was formed to help feed those in poverty-struck post-war Europe. Americans could buy the packages at stores and railway stations or through their local churches or civic groups for between $10 and $15. The packages would be sent to relatives or friends across the Atlantic, or could simply be addressed to 'a hungry person in Europe'. Each box, weighing about 30 pounds, contained enough rations to sustain one person for ten days (or ten people for a day).

One of these packages is now in the Smithsonian's National Museum of American History. It is a simple pale brown box with dog-eared corners and the words 'Care. United States of America' emblazoned on its side, along with its weight (30 pounds) and dimensions (0.613 cubic feet). Inside are seven smaller boxes and bags containing macaroni, cornmeal, instant chocolate-flavoured drink mix and non-fat dried milk. A note with it says: 'Greetings from the men of U.S.S. Lake Champlain'. At first, the contents of these packages consisted of leftover army rations stockpiled by the USA for the invasion of Japan, but when those ran out, Care produced its own, including more meat and fats in the line-up. As Europe lay in ruins with many on the verge of starvation, hundreds of thousands of Care packages were sent across the Atlantic. During the airlift, more than 200,000 of them were delivered to Berlin. With every single box purchased by individual donors such as the men of USS *Lake Champlain*, the initiative was a remarkable person-to-person relief effort.

During the airlift, the philanthropic activities of one particular individual caught the world's imagination. His name was Gail Halvorsen and, during a rare break between flights, the young American pilot from Utah wandered to the edge of Tempelhof airport and started talking to some children on the other side of the wire fence. As he turned to leave, he reached into his pocket and drew out two sticks of chewing gum. Breaking them in half, he passed the four pieces through the wire to the children who had been translating for him. Halvorsen was surprised to find that the children did not immediately start a brawl over the gum. 'They came in close, but no fighting,' he recalled in an interview with CNN in 1995. 'And then the kids that didn't get any, they wanted part of the wrapper. So the other kids tore off the outer wrapper and then the tin foil and passed it around and they smelled it, [and] wow! Their eyes got big, like they remembered when they could have gum. And I couldn't believe it. I just was amazed at what the smell of a wrapper meant.'

Halvorsen was so impressed with the children that he told them he would return the next day and, if they promised not to fight over it, he would drop more candy and gum in a parachute made from his handkerchief. But how, the children asked, would they spot him amid the streams of planes arriving at Tempelhof? Halvorsen told them he would wiggle the wings of his plane, as he used to do when flying over his parents' house back home in the USA. The next time 'Uncle Wiggly Wings', as Halvorsen would become known, arrived at Tempelhof (this time with 20,000 pounds of flour loaded in his plane), as he neared the airport he pushed the candy he had collected from his rations out of the emergency chute behind the pilot. He could not see what happened to it but, once the plane had landed, he looked across to the other side of the airfield. 'There were the kids, three arms through the barbed wire, waving the parachute at all the airplanes,' he told CNN. 'I knew then that they'd caught it.'

It was the start of 'Operation Little Vittles'. At first, things continued on an informal basis with Halvorsen and his fellow pilots collecting chocolate and gum and creating handkerchief parachutes. Initially, they dropped the candy – taken from their own personal rations – secretly, since they had no official permission to do so. But when a package almost hit a German newspaper reporter on the head, word got out about the secret mission of Halvorsen and his crew. Initially, senior officers were angry at Halvorsen's unauthorised action. It soon dawned on them, however, that this was a worthwhile enterprise bound to generate goodwill. He was told to carry on.

Then, as Halvorsen puts it, 'the floodgates opened'. The American Confectioners' Association told him they would send over all the candy he could drop. Eventually, the pilots had to stop dropping packages around Tempelhof because the crowds were becoming dangerously large. Instead, they dropped it all over West Berlin, in parks and football fields. Operation Little Vittles was now a well-established part of the airlift. Children from twenty-two schools in Chicopee, Massachusetts, made the parachutes, working in an old fire station that became Operation Little Vittles' headquarters. By January 1949, Halvorsen – who was given the nickname the 'Candy Bomber' – had released more than 250,000 parachutes loaded with candy and gum from his C-54 Skymaster.

Carrying 10-tonne loads, the four-engine C-54s were the chunky workhorse planes of the lift. They were brought in to assist the C-47 Skytrains (the Gooney Birds, as they were affectionately known) with which the operation was started. From Britain's Royal Air Force, there were aircraft such as Dakotas, Bristols, Hastings and Lincolns. The success of the airlift was to depend on these machines. Nevertheless, a certain amount of improvisation was needed to prepare the planes to deliver supplies. After all, few of them had been designed to carry cargo. At first, passenger aircraft with their seats ripped out were used. As the airlift progressed, doors were removed to speed up the unloading process.

Moreover, the shortage of planes was at times acute. According to one story, a diplomat who landed at Wiesbaden in a C-47 as part of a tour of Europe left his plane for half an hour to get a snack while the plane refuelled, and when he and his crew returned, they found it loaded up with sacks of flour. Special planes had to be used to carry salt, which ate away at the fabric of most aircraft. British Sunderland and Hythe flying boats (designed to take off and land on water and therefore treated against corrosion) were brought in. This worked well until winter, when the lakes froze, prompting another of the many ingenious improvisations generated by the airlift. Rust-proof copper panniers were attached to the bays of British bomber planes, allowing salt delivery to continue without harming the planes.

The departure point for the British and American aircraft was in West Germany – airbases such as Wiesbaden, Celle, Rhein-Main (which had originally been designed for airships such as the *Hindenburg* and the *Graf Zeppelin*) and Fassberg. Pilots would be given an exact take-off time that had to be followed to the second. After a couple of hours in the air, the planes would reach Berlin, landing at Tempelhof airport, Gatow and, later, Tegel, which was built in 1948 in less than three months to expand the capacity of the operation and was christened like no other airport – with a delivery of 20,000 pounds of cheese. Once on the ground, teams unpacked their cargo, offloading by hand at a rate of a thousand pounds a minute in an astonishing thirty-minute turnaround.

The man behind this remarkable efficiency was General William Tunner. Known as 'Willie the Whip', Tunner had a reputation for achieving the impossible. During the Second World War, he had masterminded a daring airlift over the 'Hump' of the Himalayas from India to China after the advancing Japanese cut off the Allies' transport artery, the Burma Road, in 1942. 'Tunner was a brilliant, dedicated, meticulous leader whose steel-blue eyes and index-card mind missed nothing,' writes military historian Roger Miller. 'A workaholic, he labored long hours at an intense pace and drove his staff relentlessly.' This logistics junkie had been watching the airlift from a distance and was desperate to get his teeth into it. 'I could tell, from the reports coming in, that the operation of the Airlift could stand improvement,' wrote Tunner in his memoir. He viewed newspaper reports celebrating the valour of pilots who continued to fly despite their exhaustion as a sign of poor management. 'To any of us familiar with the airlift business, some of the features of Operation Vittles which were most enthusiastically reported by the press were contraindications of efficient administration,' he noted with scorn. Tunner was itching to sort things out.

The general got his chance. On 28 July 1948, as the new head of the Airlift Task Force, he flew to West Germany. The forty-two-year-old air transport officer soon put his personal stamp on the airlift. With a handpicked staff and his personal secretary, Katie Gibson, Tunner conducted detailed studies into the pilots' downtime and work deployments. But while his version of efficiency was backed up by plenty of charts and tables, he did not shut himself away in an office. Much of his time was spent out on the tarmac or climbing into planes to talk to pilots. 'In the dark, wearing his worn flight jacket or covered with coal dust, he often appeared to be just another officer, slightly older than most,' writes Miller. This hands-on approach allowed the general to remove inefficiencies from the operation. The smallest details did not escape his eyes. Tunner noticed that while planes were being unloaded pilots would wander over to the terminal to get coffee and doughnuts before heading back to start up the engines again. This, reckoned Tunner, was precious time wasted. He introduced a mobile snack bar that would roll up to the aircraft as soon as it taxied to a halt. The same system delivered flight schedules and weather reports to the pilots.

Based on discipline and precision, the general's logistics regime employed a 'right first time' policy. Pilots had one chance to land and, if they missed it, were required to return to their base and try again. It was a radical shift from the system that had operated before Tunner's arrival – one where, in bad weather, planes were stacked at different altitudes, where they circled until it was safe to land. On 13 August 1948, it became clear that this system was untenable. Tunner describes the chaos of 'Black Friday', as it became known. 'The ceiling had suddenly fallen in on Tempelhof,' he wrote. 'The clouds dropped to the tops of the apartment buildings surrounding the field, and then they suddenly gave way in a cloudburst that obscured the runway from the tower. The radar could not penetrate the sheets of rain.' A couple of aircraft made crash landings. Meanwhile planes, still arriving at three-minute intervals, were piling up in a massive stack that soon occupied the airspace between 3,000 and 12,000 feet. It was a giant airborne traffic jam. Tunner's plane was in the middle of this mess (ironically, he was coming to Berlin to attend a ceremony honouring the efficiency of the airlift). Fearing a mass collision, he grabbed the mike and told the air traffic controllers to send every plane back to its home base. 'There was a moment of silence, then an incredulous-sounding voice said, "Please repeat." "I said: Send everybody in the stack below and above me home. Then tell me when it's O.K. to come down."'

Tunner's unexpected decision provided the basis for the new system. After Black Friday, if a pilot missed the approach to Berlin, he had to turn around and take his load back. Flying at a different altitude from the planes that were returning unloaded, the pilot would head back to his airbase, where his plane would be given a new crew and slotted back into the flow of aircraft to start all over again. Tunner had created a high-altitude conveyor belt of planes, evenly spaced, all moving at the same speed and all bringing essential goods to Berlin.

The general was not interested in daredevil flying or other acts of heroism. His methods were based on getting everybody to keep to the system. Procedures had to be followed with unerring regularity. 'This steady rhythm, constant as the jungle drums, became the trade-mark of the Berlin Airlift,' he wrote. 'I don't have much of a natural sense of rhythm, incidentally. I'm certainly no threat to Fred Astaire, and a drumstick to me is something that grows on a chicken. But when it comes to airlifts, I want rhythm.' Under Tunner's regime, planes were landing or taking off from Berlin's airfields every three minutes round the clock, carrying an average of 8,000 tonnes of provisions into the city every day. It was an industrial approach to a political battle – with moveable food at its strategic heart.

Tunner also realised that, as in any battle, keeping up the morale of the troops was essential. He encouraged competition between aircrew teams. Each airbase was given a goal and the results were posted on a large 'Howgozit' board. The *Task Force Times* kept everyone informed, reporting on the achievements of particular crews. A tonnage column developed the spirit of competition Tunner was after, providing the incentive for aircrews to beat the previous record. The *Times* was also entertaining. The radioman from Tunner's own plane, John 'Jake' Schuffert, turned out to be a brilliant cartoonist and in each edition of the paper, he poked fun at a different aspect of the airlift. One of his cartoons depicts bemused Berlin children who, amid a shower of handkerchiefs bearing chocolate and chewing gum from the Candy Bomber, spot a couple of bars floating down in a double parachute created by the cups of a woman's brassiere.

Humour kept the spirits up in what for the pilots was a dangerous and exhausting mission. Berlin's 20-mile-wide air corridors were tricky to negotiate, requiring some of the most precise, skilful flying ever seen. In the early days, pilots took risks, occasionally descending to below the 400-foot minimum before breaking out of the clouds (after Tunner took over, anyone breaking the 400-foot rule was threatened with court martial). Moreover, the aircraft were taking a hammering, both from the goods they were carrying and from the relentless turnaround times. It was inevitable that in such conditions there would be accidents. At least thirteen German civilians died during the airlift. The British lost thirty-nine airmen and civilian airlift staff and the Americans lost thirty-one men during the operation. Most of the fatalities were caused by equipment failure.

There were other, unexpected, perils, too. Coal or flour dust could explode, if it built up inside a plane. Doug Southers, then a twenty-year-old American flight engineer, remembers this threat well. Southers, who made almost 200 trips to Berlin without leaving the airfield, was mainly responsible for hauling coal, in what were known as 'gutty sacks' made of hemp, as well as flour in cotton sacks. The inside of his plane was constantly coated either with black from the coal dust, or white from the flour – dustings that were highly combustible, particularly when combined. In another of the many examples of clever improvisation, the pilots found a way to create sufficient ventilation to prevent the trapped dust from accumulating and exploding. 'We took the escape hatches out over the wings, put them in the back of the plane and we never put them in again,' Southers recalls. 'Rain or shine, we just left them open.'

For Berlin, the airlift was a curious reversal of history. For a start, it turned one of Hitler's most ambitious pieces of architecture into a strategic weapon in the fight to preserve West Berlin's freedom. Tempelhof – the most celebrated of the airfields used in the lift – had certainly not been planned with such an operation in mind (when the Americans took control of the airport, they soon slapped paint on the 20-foot eagle and superimposed the stars and stripes over the swastika on the globe clutched in its claws). Tempelhof Weltflughafen, or

'World Airport', had been destined to be Hitler's gateway to Europe. Designed by Ernst Sagebiel, construction began in 1936. Today, it is still a vivid embodiment of the Führer's mighty aspirations. From the terminal, two great curved wings housing hangars and workshops sweep out in a dramatic arc that is half a mile long. The building has only twenty per cent fewer square feet of floor space than the Pentagon. By the time the main structure was completed in 1937, it was Europe's largest building, and it is still thought to be the world's third largest by area. Tempelhof's vast spaces were conceived to welcome unprecedented numbers of people. Space around the airport was designed to accommodate 10,000 cars and the fourteen towers that at regular intervals punctuate the building's curved wings were to serve as dramatic viewing points for some of the more than 60,000 spectators Hitler envisaged would attend air shows. By 1948, the airport was welcoming visitors of a different kind – British and American pilots delivering supplies to the people of Berlin.

For Berliners, the change in allegiances must have been dramatic. After all, it was British and American bombers that had turned much of their city into rubble, and yet the same pilots were now engaged on a dangerous mission to help keep them alive. Berlin had been the capital of Hitler's Third Reich. Now western governments were spending a fortune on feeding citizens of a nation that, until just three years previously, had been a bitter wartime enemy (by the end of September 1948, the cost of the airlift had reached almost $400,000 a day).

At the same time, British and American pilots were altering their perceptions of the Germans. Doug Southers remembers the shift in attitudes brought about by the operation. 'Things really changed after the airlift started,' he recalls. 'People said that before the airlift if an American soldier went into a café, the Germans would get up and leave. A week after the airlift started, it was altogether different.' When it was announced that Germans were to be employed as mechanics on the maintenance and repair of the planes being used in the lift, some American pilots feared the Germans would try to disrupt the operation. However, nothing Southers saw led him to believe that this was their intention. 'They were excellent mechanics,' he says. 'And I was never afraid at all – I never had one feeling that the German mechanics would sabotage the planes.'

The airlift came at a difficult time for Britain. In 1948, its citizens were still living under strict rationing rules of their own (a regime that was to continue until 1954). In many ways life in Britain was harsher then than during the years of conflict. Indeed, when the 1948 Olympic Games were held in London, the British government encouraged participating nations to bring their own food with them and special dispensation was given to British athletes to consume a daily allowance of chocolate and sweets to prevent their better-fed competitors gaining the dietary advantage. Worried about privations on British shores, the Argentine team even conducted its own mini-airlift, flying in 100 tonnes of meat with its athletes. Yet over in Berlin, people were receiving butter, coffee and sugar, as well as the occasional Care package from America containing such delights as canned ham and fig pudding.

By spring 1949, the airlift was running smoothly and the daily tonnage hauled continued to rise. The allies were winning this new war with sheer dogged perseverance. They made it clear that, if necessary, the airlift could continue indefinitely. And any lingering doubts as to their resolve were buried by the performance witnessed on 15 April 1949. In an operation known as 'Easter Parade', the goal was to fly 10,000 tonnes into Berlin in twenty-four hours (fifty per cent more than the previous record). The idea was to deliver a forceful message to Stalin that the airlift was unstoppable. The teamwork seen on that day was astonishing. Freight crews challenged each other to break new records for offloading. Cigarettes were given as prizes to the winning teams. Tunner, of course, was at the scene, praising the crews as well as raising competitive spirits by spreading rumours of what rival teams were achieving. By noon on Sunday 16 April, after almost 1,400 sorties and with planes taking off or landing every thirty seconds, nearly 13,000 tonnes of supplies had been flown into Berlin, enough to fill a freight train 4 miles long. It was a masterful display of logistical expertise. On 5 May 1949, the Soviet Union agreed to lift the blockade and, at midnight on 12 May, the borders were opened.

The success of the airlift had astounded even the most optimistic of its proponents. Certainly Stalin had not predicted that food would be what upset his plans for controlling Germany. Yet the Russian leader had made a double miscalculation. Not only did he underestimate the logistical prowess of the western allies. He had failed to predict the dramatic psychological change in Berliners from conquered people to willing participants in a British–American initiative, helping build runways and offload supplies.

While the Berlin Airlift was where hostility between the Soviet bloc and the west first emerged, the co-operation of former enemies during the airlift began to heal many of the psychological wounds inflicted by the Second World War. The airlift had other achievements, too. It brought the western allies closer together (instead of driving them apart, as Stalin had hoped) and paved the way for an independent, democratic West Germany, closely tied to the west. The operation, moreover, led to the formation of the North Atlantic Treaty Organisation, or Nato, a collective European defence pact that was to unite democratic states across Europe and North America. Moveable food had proved an astonishingly powerful tool in post-war strategies.

If supplying Berlin delivered the west its first victory in the Cold War, the city was also the focus for that war's conclusion. It was the dismantling in 1989 of Berlin's famous wall (constructed in 1961 to separate the eastern sector from the west) that signalled the demise of the communist bloc. How humiliating it must have been then that, more than four decades after the end of the airlift, the Russians found themselves the recipients of food supplies. In November 1990, with the Soviet Union facing its harshest winter since the end of the Second World War, appeals for help were met by, among others, the West German government. Germany's shipments included 5,000 tonnes of dried potatoes, more than ten million cans of sardines and nine million cans of tuna fish. In a bizarre turnabout, these were supplies of food that had been stockpiled in Berlin and maintained in case the city was ever again to be besieged by the Russians.

It was an unlikely scenario. Stalin had been resoundingly defeated by the gritty resolve of the allies. He was hardly likely to try again. But while the Easter Parade had been a dazzling logistical flourish, it was the daily efforts of the pilots, plodding relentlessly on, that really broke Stalin's will. Tunner understood this. 'The actual operation of a successful airlift is about as glamorous as drops of water on stone,' he wrote. 'There's no frenzy, no flap, just the inexorable process of getting the job done.' In 1948 and 1949, that job was keeping West Berlin out of the Soviet grip. What motivated Tunner and his colleagues is powerfully visualised in a Douglas Aircraft Company poster of the time. In the poster, a young girl holds a glass of milk in her hands that is at the end of a stream of glasses of milk emanating from the belly of a plane above her. Dozens of aircraft fill the sky around. Across the poster are the words: 'Milk . . . new weapon of Democracy!' For the western allies, it was moveable food, not loaded guns that had proved the most effective piece of equipment on the newly formed Cold War battlefield.

5

Tiffin Travels

Curry catches the corporate imagination

Curry: Spicy dish from India, Thailand and other Asian countries

Origin: The term was used by the British to describe an unfamiliar set of Indian stews and ragouts

Etymology: In South Indian languages, *karil* or *kari* refers to spices for seasoning or to dishes of sautéed vegetables or meat

Legends: Some say that Chicken Tikka Masala is actually a British invention, created when a fussy English diner insisted on some gravy with his Chicken Tikka. An enterprising chef complied by adding a tin of Campbell's soup and some spices to the dish. Chicken Tikka Masala is now becoming popular in India

USINESSPEOPLE LOVE FANCY management techniques. Quality Circles, Total Quality Management, Boundarylessness and Authentic Leadership are among them. In the late 1980s, one particular technique caught the attention of the business world. From the conversations going on between executives you might have been forgiven for thinking that dozens of them had suddenly taken up karate. Terms such as 'green belt', 'black belt' and 'master black belt' peppered the office banter. However, the executives would have in fact been referring to the hottest technique then on the management market – one that, in an odd muddle of cultural references, not only draws upon eastern martial arts but also incorporates a Greek symbol into its title.

Six Sigma is a quality control technique that multinational companies such as Motorola and General Electric claim has transformed their businesses. Taking its name from sigma, the lower-case Greek letter σ, which is used as a symbol for deviation, Six Sigma is a statistical term used to measure how far a process deviates from perfection. Achieving the Six Sigma benchmark means having no more than 3.4 defects in a million products or service transactions. If that were a spelling test, it would mean making fewer than four mistakes per million words (the equivalent of twelve copies of this book). Applied to manufacturing and logistics, the system is executed by 'master black belts' – staff trained in advanced statistical techniques who spread the system to other parts of the business – and is meant to reduce defects, cut costs and improve customer satisfaction. To outsiders, however, talk of 'statistical process controls' and 'improving the process capability' can seem like meaningless gobbledygook.

As they hurry to work in Mumbai, India's chaotic commercial capital, it is unlikely that these terms enter the thoughts of one group of workers. They are too busy heaving food from place to place. Every day, dressed in white cotton kurtas and Gandhi caps, about 5,000 men (and a handful of women) turn up at households in the outer suburbs of Mumbai to collect cylindrical stacks of metal tins filled with vegetables, spicy meats, dhal, rice and chapattis. These are the lunches prepared by housewives for husbands who left for work several hours earlier. Once they have collected all their tins, the dabbawallas ('dabba' means lunchbox and 'walla' means the person associated with the trade) head down to the local railway station. There, they sort their cargoes according to destination and pack them into the luggage compartments of commuter trains. A couple of hours and several pairs of hands later, the tins of food will end up at offices, factories and schools all over the city to be consumed by hungry executives, workers and students.

In 1998, *Forbes*, the US business magazine, gave this eccentric distribution system a Six Sigma efficiency rating indicating a 99.999999 percentage of correctness, putting the thousands of workers delivering lunches on a similar efficiency footing to giant companies such as General Electric and Motorola, who are among the other recipients of the award. But unlike the logistics systems deployed by those corporations, this one uses feet, bicycles and local trains to deliver the goods. While global transport companies build up fleets of trucks and planes as well as sophisticated tracking technology, here in Mumbai no investment has been made in equipment or facilities. No databases, paperwork or computers are used to operate the service. Two dusty terminals in one of the organisation's tiny offices look as if they have not been touched for decades. Few of the employees are educated, and many are illiterate.

Yet they claim to deliver more than 170,000 meals a day across the city – all in the space of a few hours, making almost no mistakes. General Tunner would have been impressed. Every day, tens of thousands of meals cooked by wives and mothers travel from the kitchen stove to husbands and children in schools, shops, government facilities and corporate offices with almost no mix-ups. It is a remarkable process, but if that were not impressive enough, after lunch, in a rare example of reverse logistics, the empty tins will be collected and – using exactly the same process – returned to the housewives who packed them with food earlier that day.

Given the system's limited resources, such accuracy seems improbable. However, Raghunath Medge can explain it. 'Our computer is our head and our Gandhi cap is the computer cover to protect it from the sun or rain,' he says. Medge is a tall, handsome fifty-two-year-old. He has a large moustache, a broad smile and a picture of himself standing next to Prince Charles stored on his mobile phone. The photo was taken in London at the wedding of the prince to Camilla Parker Bowles, to which Medge and his colleague Sopan Mare were invited (Prince Charles had first encountered the dabbawallas on an earlier visit to Mumbai, and had clearly been impressed). Tiffin delivery is in Medge's DNA. His father and grandfather were dabbawallas, and he has been re-elected president of the lunchbox system's governing body, the Nutan Mumbai Tiffin Box Suppliers' Association, three times. It is easy to see why. He combines a playful sense of humour with an intense pride for his dabbawallas (he says he knows them all, if not by name, by face). One moment he is joking with his colleagues, the next, he is talking gravely about the importance of discipline, trust and training.

The workers whose welfare Medge looks after are also known as 'tiffin carriers'. Tiffin is an old colonial term. Often thought of as a snack taken with afternoon tea, it actually refers to a light lunch eaten at midday. Indian colonial writings make numerous references to tiffin. In an 1873 memoir of his career as an administrator in the Indian Civil Service, John Beames recalled the importance of tiffin in the daily routine: 'I wrote replies to letters from the Commissioner, Board and other officials, and was usually a good deal hindered and interrupted by Deputy Collectors and other officers coming in to speak to me about this or that. Generally, however, by two o'clock the correspondence was finished. Whether it was or not, at two we had tiffin – and we wanted it.'

Tiffin was not always the lightest of meals. In a 1904 account, Eliza Ruhamah Scidmore, an American travel writer and photographer, described the overindulgent culinary order of the day in colonial Calcutta: 'The solid two-o'clock tiffin, following the heavy ten-o'clock breakfast, is so soon succeeded by the four-o'clock tea and the eight-o'clock dinner, that it is a surprise that any one survives the constant feasting which fills Anglo-Indian life.'

Since then, the word tiffin has passed into the vocabulary of Indian English. Perks for government workers often include 'tiffin allowances', while the seventy-five-year-old Mavalli Tiffin Rooms in Bangalore is one of the country's most celebrated restaurants. The tiffin tradition has even made its way on to fashion industry runways. Among the creations of edgy Mumbai designer Aki Narula is a collection of whimsical handbags in the form of tiffin boxes.

Tiffin is not the only moveable midday meal. The sandwich – the ubiquitous lunch for office workers on the run – was devised in the eighteenth century by John Montagu, the Fourth Earl of Sandwich, when he asked to be brought some meat between two slices of bread. While legend has it that he wanted to be able to eat with one hand while playing cards with the other, the more accurate story is that, as a politician who put in long hours (he reportedly wrote seventy letters a day), he favoured a meal whose portability allowed him to continue working uninterrupted. Sandwiches are also the central element in one of the more enjoyable moveable feasts: the picnic.

Then there is the Cornish pasty, the eminently moveable lunch. This self-contained meal was once the midday food of miners in Cornwall. A sealed pie containing meat and vegetables, it was easily carried down into the mines – the pastry was the edible packaging – and eaten without utensils. Two knobs of twisted pastry at each end could be held with dirty hands (an important consideration since arsenic was present in the mines) and thrown away after the rest had been consumed. Like Mumbaikars, Cornish miners insisted on eating the pasty cooked by their wives so to prevent mix-ups, initials were pricked out with a fork or fashioned from pastry strips. Still popular among the British, the pasty occasionally crops up in unlikely places, such as Real del Monte in Mexico, where the pies – given a local twist, with pineapple versions and jalapeño on the side – are a reminder of the nineteenth-century presence of Cornish miners in Latin America.

While sandwiches and pasties are easily carried to work, curry and rice present a tougher challenge. Moreover, anyone who has witnessed the morning rush hour at Victoria Terminus, an enormous Gothic building that is Mumbai's largest commuter station, will understand why Mumbaikars rely on others to bring them their midday meal. Trains are packed with people. With doors that never close, passengers occupy space inside and outside the carriages, many clinging to the sides to secure a ride to work. This suburban railway system is one of the most densely packed commuter services on the planet. Extra carriages are reserved for women at rush hour and, with more than six million passengers a day, the service frequently experiences a 'super-dense crush load' of up to sixteen people for every 11 square feet of floor space. No wonder few commuters want to travel with tins of hot curry, rice and chapattis.

The super-dense crush load is just one ingredient in the complex mixture of social, cultural, economic and geographic factors behind the growth of the dabbawalla delivery system. Industrial expansion provided the initial catalyst. The habit of having tiffin delivered emerged in the mid-nineteenth century as a cotton boom erupted across what was then Bombay. When, in 1861, the American Civil War cut off supplies from the USA, demand in Britain for Indian cotton soared. Once the railway linked Bombay with the cotton-growing districts of central and southern India, the city became the port from which cotton was exported to Britain's textile industry.

Industry and trade attracted workers and merchants from across India. Sprawling residential areas arose to accommodate the newcomers. As domestic housing moved south, far from the textile mills and trading houses in the north, workers could no longer return home for lunch. Then, according to one story, more than a century ago, a Parsi banker (Parsis are descendants of Zoroastrians from Persia) hired an errand boy to bring lunch from his home in the Grant Road area to his office in the Ballard District. This boy picked up additional orders and quickly enrolled members of his family from their village near Pune, south-west of Bombay. Others say it was not a Parsi but a British colonial officer who decided to have lunch cooked by his wife and brought to his desk, sparking the creation of the lunchbox service.

Whichever is true, the dabbawallas still hail from a region near Pune. They speak the same language (Marathi), share the same branch of the Hindu religion, with Vithoba as their presiding deity, and claim to be descendants of the soldiers of the Maharashtrian warrior king Shivaji. Many are related to one another, creating a close-knit community of workers. Teams are overseen by 'mucadams'. Familial and regional ties create a sense of pride and strong bonds of trust between the members of each team, something seen as essential in a system whose workers have unimpeded access to offices across the city and to households where, by the time they arrive, male members are absent.

Geography still plays a role. Residential suburbs such as Andheri, Juhu and Bandra are located north of Mahim Creek, while the city's commercial districts lie far away in the south. Although these patterns are changing, the city's long, thin topography still makes returning home for lunch (as workers in the rest of India often do) impossible for most Mumbaikars.

Mumbai's multitude of religions and ethnic groups creates another compelling need for the dabbawalla system. While Delhi, with its foreign diplomats, politicians and journalists, remains the most international of Indian cities, Mumbai is a microcosm of India's domestic multiculturalism. This is partly thanks to the policies of one of Bombay's most celebrated governors, Gerald Aungier. When he took office in 1672, Aungier began devising incentives to attract skilled workers and traders, bringing Parsis, Armenians, Jews, Gujaratis and Brahmins to the city. Within a decade, Bombay's population had tripled. Workers flocked to the city to find jobs in the mills. When the cotton boom drew to a close in the late 1860s, Bombay went on to become the country's textile centre, with Parsis and Gujaratis moving in to build businesses producing cloth for domestic markets. Today, as the country's commercial capital and home to the Bollywood film industry, Mumbai still draws in migrants from across India as well as from Bangladesh, Pakistan and Nepal.

As a result, most of the world's religions have followers in Mumbai. Muslims head across a narrow causeway to the Haji Ali mosque, romantically sited on a chunk of land in Mahim Bay. Fire temples serve as the shrines for Mumbai's Parsi community. In Prabhadevi district, the enormous Shree Siddhivinayak temple is responsible for traffic chaos every Tuesday, when thousands of barefoot Hindus make their weekly pilgrimage to the shrine. Buddhists and Jews are a small but important part of the city's multicultural make-up. A handful of Roman Catholic churches are reminders of the days when the Portuguese controlled the city before foolishly handing over this valuable piece of land to the British in 1661 as part of Catherine of Braganza's dowry when she married Charles II.

With myriad religions come strict rules on eating – and dietary restrictions are taken seriously in India. When told she might die if she did not drink beef tea, Mahatma Gandhi's wife Kasturba, a Hindu, declared: 'I will not take beef tea. It is a rare thing in this world to be born as a human being, and I would far rather die in your arms than pollute my body with such abominations.' For Hindus, religion comes with a caste system that also governs eating habits. 'In no particular of their daily life are the Hindus affected more by the rules of caste than in reference to eating and cooking,' wrote the author of *The Indian Mirror* in 1878, recounting the story of an American missionary who witnessed a high-caste Hindu 'dash an earthen jar of milk upon the ground, and break it to atoms, merely because the shadow of a Pariah had fallen upon it as he passed'.

As the British colonial powers discovered to their cost, challenging the culinary divisions within Indian society can have serious consequences. The first major uprising against the British in India – the Sepoy Rebellion or Great Mutiny of 1857 – was sparked when rumours spread that cartridges for the newly issued Enfield rifles (whose wrappings had to be bitten off before loading) were greased with cow and pig fat. The move was seen as a deliberate attempt by the British to undermine the religions of Muslims, who would not eat pork, and Hindus, who would not touch beef. Of course, there was more to it than that. With the outbreak of rebellion, Indian citizens had begun to express a deeply felt resentment at the continued presence in their country of a foreign power. However, the fact that a widespread national revolt in which thousands died was triggered by the rumoured presence of animal fat on gun cartridges demonstrates the gravity with which Indians take food restrictions.

Such strong feelings persist. While the country has a secular constitution, which does not recognise caste, many still adhere to traditional caste divisions, and reinforcing these divisions is food. Hindus believe they will 'lose caste' if their food is 'contaminated' by someone of the wrong social or religious order. For many Hindus, for whom the notion of 'pure' and 'impure' still holds profound spiritual meaning, it can be of critical importance to know not only where the food came from but also how it has been prepared and who has handled the containers in which it is served.

In India, the food industry naturally treads warily. Even in the trendy All Stir Fry, an Asian-fusion restaurant in Mumbai's Colaba district, the menu's author has been careful to point out that 'separate woks and utensils are used to prepare vegetarian meals'. Foreign food companies are even more cautious. Before McDonald's opened its first restaurant in India in 1996, the company spent six years studying local tastes. It came up with the Chicken Maharaja Mac – two grilled spicy chicken patties in buns with cheese, a margarita spread (a blend of vegetable oil, egg powder, tomato, sweet relish, onion, garlic, chicken powder, yeast extract, vinegar and herbs and spices), shredded onions and lettuce and two slices of tomato. The McVeggie burger and the McAloo Tikki (a spicy potato dish) also made their way on to the menu. In India, McDonald's mayonnaise is made without eggs and different sections of the kitchen are used to cook vegetarian items. Young Indians may enjoy a McDonald's because it is 'western' (a word that for them conjures up everything that is fashionable). However, because the burger is commonly associated with ham and beef, it is still treated with suspicion by many Muslims and Hindus. Even when religion is not a consideration, Indians tend to see restaurants as expensive and unhealthy ('too oily' is the frequent complaint). In short, most Mumbaikars still prefer to rely on a home-cooked meal delivered by a reliable man in a Gandhi cap.

Logistically, the critical element in the tiffin system's success is Mumbai's suburban railway network, an extensive system that was built up from the 1850s by the Bombay, Baroda & Central Indian Railway Company. A breakdown of the local railways is one of the few things that can prevent the delivery of dabbas and part of the monthly dues dabbawalla groups pay to the Tiffin Box Suppliers' Association goes towards securing space in the luggage carriages of commuter trains. The network provides a reliable transport infrastructure whose routes, because they are designed to convey workers from their homes to their workplaces, match exactly the needs of the dabba-wallas delivering their lunches.

At about 9.30 a.m., the system gets going. In kitchens around the city, women hastily finish preparing meat curry and rice or dhal and vegetables. As they put the last touches to the meal, their local tiffin man arrives at the door, takes off his shoes and waits patiently in the entry hall for his client to bring out the daily stack of tins. He then heads back down the corridor and out into the street. Black and yellow auto-rickshaws buzz past like giant motorised bumblebees with 'Horn OK please' and 'Safety First' painted on their rears. Women in saris ride side-saddle on motorbikes beside large sports utility vehicles with smoked glass windows. Traffic manoeuvres past al fresco hair salons that share space with herds of goats. Art deco apartment blocks and crumbling Victorian Gothic mansions sit in the shadow of glittering tower blocks and hoardings advertising mutual funds, Japanese cars and satellite radio stations. Remnants of the country's colonial past are disappearing beneath trappings of the new India.

Curiously, the dabbawalla system is not among modernisation's casualties. Medge claims that customer numbers are rising by ten per cent a year, and certainly the morning activity out on the streets seems to point to a thriving business. By about 10 a.m., dabbawallas are arriving at suburban railway stations. Armies of bicycles have been parked and the collection of tins and bags hanging from the handlebars are being unloaded. Soon, tiffin boxes are scattered across the platform. Everything seems to be in disarray, but the dabbawallas know exactly what they are doing. They work swiftly to sort the boxes and pack them into long wooden trays in time to catch the 10.30 a.m. downtown express. With a screech and a hiss, the train pulls into the station and the men leap into action. They heave trays weighing up to 100 pounds on to their colleagues' heads before racing down the platform to the luggage compartment.

With a dozen or so men and their trays safely stashed on board, the train pulls out of the station. For the next few minutes, the dabbawallas can relax. Rattling along with the doors wide open and a cool breeze blowing through, the luggage compartment is a fine place to be. The atmosphere is lively as the men chatter noisily, waving excitedly when colleagues pass by in other trains. At one stop, a man gets into the compartment with huge bags containing bright pink candyfloss that virtually fill the space. No one seems to mind, and the dabbawallas cheerfully accommodate the new arrival. This is a good time to catch up on news and gossip. One fellow has made it into a district-level cricket team and is recounting the details of a recent match. Another has announced that he will be joining the almost 30,000 people lining up to run the Mumbai Marathon on Sunday. Suddenly the dabbawallas are no longer concentrating all their energies on accuracy and speed – they are telling stories and laughing at each other's jokes.

The moment of relaxation is short-lived. As the train pulls to a standstill, dabbawallas spill out on to the platform to resume the serious business of food delivery. Outside the station, trays are laid out on the pavement and the dabbawallas rush off in all directions with tins hanging from bicycles, packed into small carts or simply carried by hand. It looks like chaos, but it is most definitely organised. A roughly scrawled set of letters on top of the tin guides each lunch on its journey. Take, for example, the mark K-BO-10-19/A/15. K would be the identity letter of the dabbawalla, BO would mean Borivili (the district from which the lunch is to be collected). The remaining marks refer to the destination of the tin, with 10 for Nariman Point and 19/A/15 indicating the fifteenth floor of the nineteenth building in the Nariman Point area. This system, says Medge, is roughly forty years old and replaced an earlier one using symbols that, as demand grew, proved insufficiently detailed.

Each sector in the system is served by about twenty-five dabbawallas, each of whom will collect up to forty tiffin boxes. Before being loaded on to the trains, boxes are sorted according to the area to which they will be delivered. At the destination railway station (some may have changed trains along the way), another group of dabbawallas sorts the tins again according to the buildings or areas for which they are destined. Individual dabbawallas then head off in the final stage of the delivery. It is a sort of massive relay race during which each lunch box will be handled by three to four pairs of hands during its journey from home to the office and then back home again – and all for just 250 rupees (about $5) a month.

The tiffin delivery service is slowly moving into the modern era, in some respects at least. A dabbawalla website was recently established, as well as an SMS service. With more women working and unable to prepare their husband's or children's tiffin box, catering companies are filling the gap. As they do so, the Mumbai dabbawalla business model has echoes elsewhere, as the quest for convenience shapes eating habits.

Like gardening or sailing, cooking is becoming a hobby – a leisure activity, not a mechanism for survival. In the USA, only fifty-eight per cent of all main dinner dishes are now cooked from scratch, according to the NPD Group, which has been tracking American eating habits for two decades. And for those who cannot be bothered to cook, the plethora of convenience foods means there is plenty to choose from.

The modern consumer love affair with prepared food was born as the TV dinner. Since C.A. Swanson & Sons sold the first one in 1953, this meal has become an icon of modern life (one of Swanson's aluminium trays is displayed in the Smithsonian Institution). The business began because the company needed to dispose of thousands of tonnes of unsold turkey after a poor Thanksgiving season. A salesman at the company, Gerald Thomas, remembered having seen small aluminium trays of food on airlines and came up with the idea of producing something similar.

Swanson thought it was a splendid idea. The company devised a turkey dinner, served in small trays, packed into boxes and frozen. In each tray were several compartments, one for the meat (initially turkey, but this was later expanded to include dishes such as chicken, beef and fish), and others for vegetables and sauces. It was the first frozen pre-prepared meal and it quickly took off. As well as the convenience, what really turned the American public on to the meals was the name – 'TV Dinners' – and packaging in a box that mimicked a television screen. After all, this was the early 1950s, and the television was still modern and chic. A year after the company launched the meals, more than twenty-five million TV dinners had been sold. Frozen dinners have since become a way of life for many families, with annual US supermarket sales of frozen meals now reaching more than $5.9 billion. In the UK too, consumers happily tuck in to 400,000 tonnes of pre-prepared meals a year.

For most people, the pre-prepared meal is something picked up in the supermarket on the way home. However, you can always have something delivered. One meal that is synonymous with home delivery is the pizza, and when it comes to pizza delivery one of the world's giant brands is Domino's Pizza Inc. In 2005, the company sold 400 million pizzas. Yet, like the TV dinner, this mammoth business grew from humble beginnings. In 1960, with a loan of $500, Tom Monaghan and his brother James purchased a small pizza store in Ypsilanti, Michigan and named it DomiNick's. The early days were rocky. As well as a sizeable debt, the brothers inherited a contract for an advertisement in the *Yellow Pages* and, because they could not afford the ad, the phone company refused to connect the telephone – leaving the business without its most vital link to customers for a month, until the brothers saw the dent this was making in their income and reluctantly paid for the ad. The following year, James decided he would rather have a car than a pizza business and sold his half of the operation to Tom to buy a Volkswagen Beetle.

It is not hard to guess who made the better investment

decision. By 1967, Monaghan had renamed the company and turned it into a franchise business, which gives entrepreneurs licences to run the restaurants in return for a fee and a percentage of the profits. Despite coming close to bankruptcy in the early days, the franchise system allowed the company to expand extremely rapidly. Between 1979 and 1985, stores were opening at a rate of just over one a day. Today, from more than 8,000 outlets around the world, this giant company sells pizzas to households in more than fifty countries. A 'College of Pizzarology', set up in 1973, trains new franchisees, and a set of thick manuals guides each store on the exact procedure to be used for everything from site selection to ordering supplies and delivery of the finished pizza.

Over the years, Domino's has come up with a number of devices that improve the way it delivers its products. Monaghan himself devised a box that would not collapse and squash the pizzas when stacked. In 1998, the company introduced bags containing a patented heating mechanism that keeps the pizza as hot as if it were in an oven during the journey from the shop to the customer's front door. Special insulation cuts out moisture, keeping the pizzas crisp. But the key to the success of the business lies in Domino's distribution system. From industrial-sized food facilities, vast fleets of trucks shunt millions of tonnes of pizza dough, cheese, pepperoni and other ingredients to Domino's outlets every day. Preparation at the store is minimal. 'Stores don't have to worry about anything but selling pizza,' potential franchisees are told on the website by Mike Soignet (his title is 'Executive Vice President of Maintain High Standards'). Domino's has even given the TV dinner a new twist. In the UK and Ireland viewers can buy pizzas via interactive television.

Collectively, Domino's pizzas cover extraordinary distances. In the USA alone, every week pizza drivers travel 9 million miles (more than thirty-seven trips to the moon and back). Yet with so many outlets, the distance of a meal from its origin to the customer's front door is relatively short. Pizza journeys are a mere hop compared to the peregrinations of a growing army of mail order dinners. In America, where families often live great distances apart and jobs take people far from their birthplace, a taste of home regularly arrives in a couriered package that has travelled thousands of miles. Shipping food long distance is also easier in the USA than elsewhere, where differing national safety regulations make the process a complex one.

From Memphis, a company called Corky's BBQ sends out packages containing full meals – pork ribs slow cooked in hickory chips and charcoal, complete with barbecue sauce, baked beans and pecan pie. From Boston, Legal Sea Foods ships lobster, crab cakes and New England clam chowder, along with portions of Boston cream pie (the company's first mail order came in the 1970s from actress Carol Channing, who was instructed by her doctor to eat swordfish seven times a day). Even American dieters – and celebrities such as Kevin Costner, Nancy Sinatra and Liza Minnelli are among them – have food couriered to their doorstep by ZoneChefs, which delivers its gourmet microwaveable meals around the USA.

During the holiday season, companies like these can barely keep up with demand. In late November, in the run-up to Thanksgiving, turkeys in their thousands fly around the country – powered by jet engines, not wings and feathers. And while many of them require cooking (for which turkey preparation hotline advisers offer long-distance assistance), fully cooked birds with all the trimmings are the increasingly popular mail order version.

Billy Howell knows all about this. Originally from New Orleans, he started cooking Cajun-style fried turkeys for family and friends in the early 1990s from his home in Dallas. His recipe was so popular, friends told him he ought to start selling them commercially. Howell realised they were right. After putting a small ad in the *Dallas Morning News* in 1993, he received more than 400 orders. He cooked them all in his garage between October and January. Today, the Cajun Turkey Company has a large factory in West Texas that cooks turkeys to Howell's specifications and packages them in his wrap ready for shipping, along with side dishes and desserts such as a bread pudding with rum sauce from New Orleans. 'People just don't cook any more – they don't have time,' says Howell. 'And I will say this about the American public – they wait till the last minute. They know that Christmas is going to be on December the twenty-fifth and they know that Thanksgiving is the last Thursday of November, but if I told you how many phone calls we get on the last Monday and Tuesday before the main event you wouldn't believe me. It's crazy, we make our year in eight weeks.'

What families may not contemplate as they tuck into their Cajun holiday turkey is that the meal has been on an extraordinary journey. Corky's BBQ, Legal Sea Foods and the Cajun Turkey Company are among many companies that use FedEx to get their meals to customers across the USA. Their packages are part of a mammoth daily operation in which about six million individual parcels are sent by planes and trucks across the USA and the rest of the world. At least 1.5 million of them go through FedEx's Memphis hub, a giant facility that is responsible for turning Memphis into the world's busiest cargo airport.

Before setting out, the cooked turkeys and pork ribs have had their journeys carefully planned. Computers map out the most efficient route for the packages. FedEx calls this 'system form'. At the Global Operations Control Centre, the company's central nervous system, a large screen displays the exact position of every FedEx aircraft in the sky. Televisions tuned to CNN and Fox News help alert teams to any natural or manmade disasters that could affect the day's deliveries. Fifteen meteorologists (two of them avid storm chasers) keep the fleet out of climatic trouble. As well as anticipating the storms likely to hit certain airports, they monitor the progress of other events, such as erupting volcanoes whose hot ash could wreck an aircraft's engine. In a global teleconference call every morning, FedEx employees go over anything that went wrong the previous day and discuss any problems that the next twenty-four hours might bring – whether that is renewed conflict in the Middle East or a hurricane in Central America. Staff at the control centre constantly alter flight plans to cope with the day's events, rerouting planes where necessary or offloading goods on to trucks. Fortunately packages, unlike passengers, do not complain about this.

If FedEx deploys hundreds of large airborne machines to deliver the goods, its business is equally dependent on a tiny rectangle of thin black lines: the barcode. The barcode on each domestic package has been scanned up to twelve times before the parcel reaches its destination. Those travelling internationally are scanned up to twenty-three times during their journey. Crucially, FedEx pioneered a method of using these barcodes on shipping labels that allowed packages to be tracked. It also developed handheld scanners to read the codes and send them back to a central computer. John Dunavant, who runs the company's Global Operations Control Centre, remembers how it all worked before the advent of barcodes. In those days, he says, couriers would have to look up the destination zip code in a fifty-page directory. 'If you had a package with a New York City zip, the courier would look it up and pick up a black marker and put EWR on the package,' he explains. 'The problem was you had a lot of option for errors – and everyone has different handwriting, especially me. It looks like I write with my foot.' To save time, couriers started memorising the directory. Dunavant, who was then a checker in the hub, had to remember the zip codes for the whole Boston area. 'We probably memorised three or four hundred zip codes. I still remember some of them: 01209 is State Street.'

Late at night at the Memphis hub, the scale of FedEx's operations becomes apparent. By about 10.30 p.m., vehicles of all kinds start converging on the airport. Trucks roll in from service stations loaded with large metal containers full of packages. Transport buses packed with workers trundle in (with 8,000 people a night showing up for work, there is not enough parking space to accommodate more than a fraction of them). Then there are the aircraft. Almost 700 planes – from wide-bodied jets to smaller feeder aircraft – fly FedEx packages round the world. More than 140 of them pass through Memphis airport every night in the space of just a few hours.

The corporate branding of all these vehicles is impressive. From workforce bus services to DC-10s, everything is decked out in the famous orange and purple, with the FedEx logo brandished across the larger surfaces (in a clever piece of design, the logo uses the negative space between the E and the

X to create a forward-facing arrow – something not always spotted by customers). Even in dingy corridors, where you might think no one would notice, metal railings are carefully colour co-ordinated in the correct shade of orange and purple.

Before starting work, 6,000 people must pass through a set of X-ray machines. Most of them will not be flying anywhere. Yet airport-style security is strictly enforced. Once inside, they gather in teams to conduct the flex and stretch exercises the company insists on before anyone starts lifting things. All sorts of machines are at the ready. There are dollies (rectangular trailers pulled by tugs used to move containers full of packages around), slave pallets (free-standing platforms for loading and unloading aircraft) and yard mules (diesel-powered trucks that transport containers full of packages to waiting aircraft), as well as 300 miles of conveyor belts. Packages on planes and trucks are now heading to Memphis from all over the country. The smell of cardboard mixes with the potent aroma of jet fuel. Everyone is gearing up for the night's big lift.

While aircraft descend on the Memphis hub, the trucks are arriving. By about 11 p.m., they are lined up at unloading bays, ready to be emptied of their contents. Positioned at right angles to a large conveyor belt, an 'extendo' (another, extendable conveyor belt) pokes like a monstrous black rubber tongue into the vehicle's interior. Two workers inside heave boxes and parcels on to it. As packages stream out of the open door and the pile inside shrinks, the extendo moves further into the truck's interior until all the packages are out and the extendo retracts to its original size. Once on the main conveyor belt, the packages move at an impressive speed. Looking down at them from the metal walkway directly above the belt, the impression is of a fast flowing river of brown cardboard. This river of

tiffin travels

packages sweeps swiftly up a sort of reverse helter-skelter to higher ground – and this is crucial, for gravity plays an important role in the trickle-down sorting process.

By now, the packages are at their highest point. They are about to enter 'the Matrix' (FedEx claims to have given its central sorting area this name before the Keanu Reeves movie came out). This is the moment where our turkeys and ribs cease to be inbound cargo and become outbound cargo. It is also where some of the most sophisticated automatic sorting machinery can be seen in action. Technology drives everything here – nothing is left to chance. Packages come swirling down the other side of the helter-skelter and slide down a massive slanting steel shelf, helped along by large arms that swing out and nudge the packages gently on to the slope. Below, workers transfer them on to one of forty-three conveyor belts that run parallel to each other. The din is tremendous as, with a banging and hissing, yellow mechanical arms shoot in and out from the sides of the belts, pushing packages off in different directions, down yet more conveyor belts to be sorted yet again.

Before they do so, these busy steel arms need instructions, since each is responsible for hitting packages destined for a different part of the world. The instructions come from a set of scanners at the head of the conveyor – the brains of the belt. As it enters the system, a criss-cross set of laser beams measures the exact dimensions of the package so that the arm, when it shoots out, can hit it bang in the centre, rather than on a corner, which would simply send it spinning aimlessly round. For each arm to identify the right parcels to punch, the boxes must be travelling a precise distance apart. Another set of laser beams therefore measures the space between parcels as they move on to the conveyor. If two are clustered together, the system will hold one back for a few seconds until it is travelling at the correct distance from its neighbour. Finally, the barcode is scanned, telling the arm down the line that a package for its region is heading its way. As soon as the package reaches the arm, bam! The yellow steel shoots out and the package heads off in another direction. More than 160,000 packages an hour travel through the Matrix this way.

FedEx, of course, ships everything from legal documents to live animals. The company once transported two Chinese giant pandas from Beijing to the Memphis Zoo (the creatures were not, happily, required to pass through the Matrix). However, food represents an increasingly large chunk of the business. Every year, FedEx handles more than thirty million food shipments, which is about 120,000 food shipments on any given business day. Among them are the cooked turkeys and grilled ribs that end up anywhere from California beaches to Wall Street trading houses.

Appropriately enough, the company's founder, Fred Smith, is the son of a wealthy businessman who made his money in food and transport (through a fast-food chain and a bus company he eventually sold to the Greyhound bus line). The young Smith came up with the idea for a courier company in 1965 in a hastily prepared term paper at Yale in which he imagined a service transporting high-value, time-sensitive goods such as medicines and computer parts by air. The Yale student was famously given a rather mediocre grade for his paper, and it was not until the 1970s, after hiring two consulting companies – at a cost of $75,000 each – to analyse the potential market and estimate the investment needed that Smith decided to go ahead with the idea.

What distinguished Smith's strategy from that of other

courier companies was the decision to have his own fleet of planes and trucks, to provide overnight services between American cities and – the key to the business's efficiency – to use a hub-and-spoke system, rather than direct flights between different cities. The term is inspired by the bicycle wheel whose metal spokes converge at the centre, and the system works by sending packages to a number of sorting locations from which they are redistributed. Even a package going to a relatively near destination first has to pass through a sorting centre. FedEx has several of these centres around the USA in cities such as Newark, Indianapolis, Oakland, California and Alliance, Texas, although these are dwarfed by the activities at Memphis, where it employs more than 30,000 people.

The hub-and-spoke concept has since become a model for all kinds of organisations. As well as its efficiency, the system (also used by commercial airlines) allows companies to get more packages (or passengers) on each flight than they can on direct flights because the system can connect small communities that would not otherwise be able to access the direct flights. Adding new destinations or pick-up points is a relatively simple matter since it requires adding only a single link back to one of the hubs.

While Smith has compared his business with the nineteenth-century clipper ships, in many ways what FedEx does more closely resembles Mumbai's tiffin delivery system. Like the dabbawallas, FedEx routes its packages through a series of centralised sorting points from where they are sent out again to their final destination. Like FedEx, the dabbawallas can reel off impressive numbers. Indeed, in numerical terms, their claim to 175,000 daily deliveries beats FedEx's estimate for its food shipments by at least 50,000.

Making sure goods are delivered efficiently to their customers is something to which company strategists devote plenty of time and energy. In corporate circles, the process is known as 'supply chain management' – a term that covers everything from sourcing raw materials to getting finished products to customers. The more effective this process, the fewer customers who complain and the fewer goods that are left lying around in a warehouse costing money. Indeed, for many companies, dull-sounding things like transport, logistics and warehousing are carefully managed elements of a business strategy. A company's supply chain management technique is a closely guarded corporate secret and software designers are constantly under pressure to devise ever more sophisticated technologies to manage the process. As a result, the careers of corporate chiefs such as Tom Monaghan and Fred Smith are scrutinised in business literature, which is gobbled up by those anxious to discover the ingredients behind their success. Six Sigma targets are treated with intense seriousness.

It is perhaps natural, then, that Mumbai's system of uneducated lunch runners has been attracting attention, and not only from British royalty. A few years ago, Richard Branson, the unorthodox chief of the Virgin Group, got a first-hand view of the service when he donned a white Gandhi cap and boarded a local train to watch the proceedings. With their trays and bicycles, the dabbawallas have charmed the corporate world. In addition to the celebrated *Forbes* magazine accolade, the system also attracted the attention of the *Harvard Business Review*, which devoted a twenty-two-page feature to the dabbawallas. Duke Corporate Education, part of a leading US business school, recently sent a dozen western bankers to follow the dabbawallas around Mumbai in the sticky heat as part of a course designed to give them a new perspective on customer service.

In their homeland, too, big local companies are taking note of the dabbawallas' success. At Mumbai's luxurious Taj Mahal – a Victorian hotel owned by India's huge Tata Group – a section of the restaurant menu is devoted to 'Bombay Tiffin', and includes dishes such as Flower Vatana Nu Ussal (a Gujarati favourite of cauliflower and fresh green peas cooked in Maharashtrian spices) and Salli Margi Ma Zardaloo (a Parsi dish of chicken cooked with apricots and topped with potato crisps). Hindustan Lever, one of India's most prestigious companies, is another fan of the dabbawallas. The consumer goods company employs 36,000 people, many of them scientists and technologists with post-doctoral degrees. The company deploys high-tech research and development methods and modern management techniques. Yet every year, it sends a handful of managers to spend a week rattling around on commuter trains with the dabbawallas in the hope that, along the way, they will soak up some of the teamwork spirit.

While he barely speaks English, Medge himself has become a popular figure at business conferences, appearing at events with titles like 'Impeccable Logistics and Supply Chain Management' and 'Breaking the Barriers'. He has made presentations to institutions such as India's Strategic Communication for Management and the Confederation of Indian Industries. In an attempt to get to the bottom of the dabbawallas' low-tech/high-efficiency model, academic studies have been conducted on the system. Among them are papers by Natarajan Balakrishnan and Chung-Piaw Teo of Singapore's National University and Joseph Antony and Rohith Somasundaram of Mumbai's National Institute of

Industrial Engineering, who include complex flow charts and headings such as 'Managed Process Links' in their study, 'Benchmarking Supply Chain Framework'.

Of course, as lunchtime approaches, questions of supply chain management, system form and logistics methodology are far from the consciousness of most Mumbaikars. Like anyone who has paid for a meal to be delivered without thinking too hard about how exactly it gets to them, the city's office workers, teachers, students and civil servants are simply wondering how soon their dabbawalla will turn up with a set of tins containing a nice bit of home-cooked tiffin.

6

Yes, We Can Do Bananas

Refrigerated ships shape Central American regimes

Banana: Fruit (technically a false berry) of a perennial tree-sized herbaceous plant

Origin: South-east Asia

Etymology: Taken by the Spanish or Portuguese from a West African word, possibly Wolof *baanana*

Legends: In a Hawaiian story, Kukali saves a valley of starving people from a vicious man-eating bird-god with the assistance of a magic banana skin. The skin, given to Kukali by his father, replenishes the fruit inside each time it is eaten. Once nourished and strengthened by the banana, the people rise up and kill their winged attacker

*A*T FIRST GLANCE, Jacobo Arbenz Guzmán and Captain Lorenzo Dow Baker appear to have little in common. One was a twentieth-century Guatemalan leader, the other a nineteenth-century sea captain from Massachusetts. Arbenz was born in 1913. He was the son of a Ladino woman and a Swiss pharmacist living in Guatemala, a country where everything from craggy volcanoes to tropical lowlands is crammed into a slice of the narrow isthmus connecting the two Americas. At the time Arbenz was born, much of his country was in the hands of foreign investors or wealthy families of Spanish descent. Most of the population were peasants living in dire poverty. Baker was born in 1840 in the pretty coastal town of Wellfleet, not far from Boston. Then, Wellfleet was a flourishing fishing community and America was a young nation whose industrial might and geopolitical influence were rapidly growing.

Arbenz was a promising child. He wanted to become a scientist or an engineer, but when his father committed suicide, leaving the family penniless, he was forced to pursue his education at the military academy. Baker also lost a parent when, at the age of six, his mother died. But while Arbenz came from an educated, middle-class Guatemalan family, Baker's father was a humble fisherman. Following in his father's footsteps, the ten-year-old Baker apprenticed himself to a fishing captain, securing a job as a cook on a fishing schooner at the age of fifteen.

The careers of Arbenz and Baker could hardly have taken more different directions – Arbenz's as an idealistic politician with hopes of instigating social reform in Central America;

Baker's as an entrepreneur. At the age of twenty-six, Arbenz married María Cristina Villanova, an educated intellectual from a wealthy family in El Salvador. Encouraged by his wife, he became passionate about eradicating Guatemala's poverty and social inequalities. Baker married Martha, his childhood sweetheart, and they had four children. As Guatemala's minister of defence Arbenz campaigned for unionisation and land reform. He and his wife mixed with labour leaders, politicians and members of Guatemala City's intellectual elite. In 1951, he became Guatemala's second democratically elected president, a leader determined to improve the lot of the poor. Baker founded the Boston Fruit Company, an enterprise that would become the multi-billion-dollar corporation that is now called Chiquita.

Despite their different backgrounds, a powerful thread runs through the lives of these two men – a thread that might never have been established were it not for the invention of a crucial transport technology: the refrigerated steamship. Refrigerated vessels delivered bananas. In a matter of decades, they had helped turn what in the 1890s was an exotic curiosity into a mass-market product, paving the way for a massive and highly lucrative trans-American trade. After all, it was one thing to persuade a sceptical American public that a yellow-skinned fruit with an embarrassingly phallic form was actually worth eating. It was another to get it to them before it had turned into a sticky pulp. Without refrigerated transport, shiploads of fruit frequently arrived at best over-ripened, at worst in a downright rotten state, making the mass marketing of bananas impossible. With the preservative power of refrigeration and the speed of steam-powered engines, however, bananas could be shipped in enormous volumes.

Bananas gave American fruit companies immense economic influence in Central American countries, allowing them to manipulate corrupt puppet regimes to suit their commercial aims – giving those countries the dubious collective nickname 'Banana Republics'. The banana boats were an essential part of the process, fuelling the growth of huge global companies such as United Fruit. Once bananas could be easily delivered to American consumers, they would transform the political landscapes of countries such as Costa Rica, Nicaragua, Colombia, Guatemala, Honduras and Panama.

While the American fruit companies extended their grip on the banana lands of these countries, America was extending its influence overseas. Between the 1890s and 1905, the USA had overtaken Britain as the world's leading industrial nation. Its companies were transforming themselves into giant corporations with investments overseas and its government was starting to intervene in world affairs.

Latin American and Caribbean countries were among the first to feel the effects of the new American imperialism, but the action was not purely military. In a shift from direct control to a more subtle form of colonialism, American banks were offering loans to regimes of strategic importance to the USA. American companies were acquiring large chunks of land overseas. It was the era of 'dollar diplomacy', a policy initiated by Theodore Roosevelt and developed by William Howard Taft and his secretary of state, Philander Knox. US corporations were considered an important element in foreign policy – and down in Central America, the companies behind the journeys of the banana were to be a crucial part of the strategy.

The rise of what is now the world's most exported fruit started with the voyage of a schooner called the *Telegraph* from Port Antonio in Jamaica to Jersey City. The *Telegraph* belonged to Captain Lorenzo Dow Baker. In 1870 Baker had filled a contract, so one story goes, to deliver machinery and supplies to gold miners working on the Orinoco River in Venezuela. On the way home, he made a port call in Jamaica. Scouting around the docks for some cargo to take home with him, Baker did not immediately see anything he fancied. Then he spotted a cluster of traders hawking unripe bananas. They seemed to be selling well, so Baker took a chance. He bought 160 bunches, paying a shilling a bunch.

Today, we take the banana for granted but to Baker it looked bizarre. In fact, it is not a fruit at all, but grows on a giant herb whose rhizome, or underground stem, sprouts something that looks like a trunk – a pseudostem that can grow up to twenty feet high. The pseudostem produces just one bunch of fruit and then dies, to be replaced by other stems that force their way up through the earth from the underground rhizome. The fruit is formed from a vast purple flower, whose weight forces the trunk to gradually bend as the oversized petals are shed and a bunch of ripening bananas emerges. This is an odd arrangement. In spite of its sexual shape, the banana is not a sexual plant – it has no seeds.

Of course, Baker did not know any of this. As he gazed at the unfamiliar green fingers on the docks at Port Antonio, his thoughts were on whether or not he might be able to sell this unknown fruit. The gamble paid off. When he docked in New Jersey a couple of weeks later the bananas, which had been stowed on deck, had transformed. In the hot tropical sun, their bright green skin had turned to a deep golden yellow and the flesh inside was soft and sweet. Baker sold each bunch for

$2. As he watched slightly bemused but eager buyers snap up his bananas, he saw more than a ripened yellow tropical fruit on a chunky stem – he saw a fortune for the taking.

Baker started bringing regular shipments of banana bunches with him on his journeys from Jamaica to the USA. He sold his consignments through a Boston-based fruit brokerage, Seaverns & Company. There, he encountered Andrew Preston, a clerk who had risen through the company ranks. Preston's quiet, unassuming demeanour concealed an ambitious and highly motivated personality ideally suited to the job of a salesman. Baker reckoned Preston was just the man he needed to help expand the business. He decided to form a partnership with him. As well as complementary personal traits, the two businessmen had assets that were well matched. Baker was by then based in Jamaica, from where he supervised the sourcing and shipment of bananas. Having founded the successful Standard Steam Navigation Company, he had the means of transportation at his disposal.

Back in Boston, Preston had the contacts and the distribution networks through which to sell the fruit. He was a tenacious businessman. Amid numerous start-up difficulties (banana volumes were difficult to predict and many perished on the voyage home) the former fruit agent's enthusiasm remained undampened. However, it was his skill as a negotiator that proved one of the fledgling company's most valuable resources. According to Charles Morrow Wilson, writing in 1941, 'the unofficious Mr. Preston demonstrated admirable skill for inspiring confidence. His quiet enthusiasms were definitely contagious and his promises were uniformly good'. The new joint venture, established in 1885, was called the Boston Fruit Company.

Meanwhile, Americans were acquiring a taste for bananas. While in the late nineteenth century, the banana was a luxury most people could not afford, this was rapidly changing. By the late 1890s, fruit companies operating in Central America were shipping millions of stems to US shores. In 1905, Americans ate an average of about forty a year. That year, the *Scientific American* – which only six years earlier had deemed it necessary to publish directions on how to peel a banana – was able to report that the fruit was 'now so abundant and cheap as to be a common article of commerce in every corner grocery store, while in the cities it is frequently referred to as the poor man's fruit'. It was the first manifestation of our voracious appetite for fruits grown in far-off countries. Today we eat kiwi fruits from New Zealand, mangoes from India and passion fruit from Thailand, often without questioning how they got to our tables. By the turn of the twentieth century, Americans were starting to treat bananas in the same way.

The banana's rise to prominence had taken an extra-ordinarily short time. 'The whole [banana] business is the development of a few decades and people still young can remember when bananas were sold, each wrapped in tissue paper, for five or ten cents, while today ten or fifteen cents a dozen is a fair price,' wrote Willis Abbot in his 1913 illustrated book *Panama and the Canal*. For American fruit companies, the banana trade was a lucrative business. It was a gold rush of sorts, but the gold was organic, not mineral, and the rush was not westwards to California but southwards to the under-developed countries of Central America. In a 1909 *New York Times* article, Frederick Palmer described the voyages of bananas in what was by then a thriving trade: 'Day after day, under the frying sun, year in and year out, the little engines of the "banana railroad" sing their chuk-chuk in the still, hot air among the motionless leaves, onward to the pier, where the Jamaican yells and sings and giggles as he starts the bunches on their journey to the pushcarts and the country grocery.'

Between 1885 and 1899, in the banana madness, more than a hundred American fruit companies were incorporated (many would later disappear). A year before the century's end, the biggest *frutera* of the lot was established, and it was an enterprise that grew out of the Boston Fruit Company. Preston and Baker's enterprise had been marching across the countries of Central America in search of potential banana plantations. Then a Brooklyn-born businessman and adventurer named Minor Keith entered the scene. Keith was a railroad builder based in Costa Rica who had been shipping bananas to New Orleans. Trouble with his New York banks and the loss of his distribution agent had prompted him to look for a new set of business associates. He quickly saw the benefits of forming an alliance with Preston, who by then was president of the Boston Fruit Company. The new enterprise, established in 1899, was called the United Fruit Company.

Keith was a dynamo – a vital force in the development of United Fruit, and a counterbalance to Preston's cool, calculating approach to salesmanship. A daring and determined character, Keith was prepared to battle tropical humidity, torrential rains, yellow fever and malaria to get what he wanted. In 1883, he had even married the former Costa Rican president's daughter, Cristina Castro, a marriage that no doubt served more than romantic interests. 'He was the sort of American entrepreneur who took his own risks, met his difficulties and dangers without asking his government for help, but claimed as his natural right whatever profits came from his enterprise,' William Franklin Sands, an American diplomat who had known Keith, recalled in a 1944 memoir. By then, that enterprise had grown into what an elderly State Department official told Sands was a 'strangling octopus' with 'the whole of Central America in its tentacles'. It was a metaphor frequently used as United Fruit's commercial network reached ever more remote corners of the region.

From the outset, the company had a powerful presence in Central America. Between them, Keith and Preston had more than 200,000 acres of land at their disposal as well as 112 miles of railroad. By the second year of its incarnation as United Fruit, the company had managed to secure a seventy-five per cent share of the twenty million banana stems then being sold in Europe and America. United Fruit did have competitors – companies such as Standard Fruit, founded by the three Sicilian Vaccaro brothers, and the Cuyamel Fruit Company, run by Russian-born Samuel Zemurray ('Sam the Banana Man', who would later become head of United Fruit).

However, United Fruit would dominate the industry in the decades to come. A 1958 study of the company by the US National Planning Association, a Washington policy research institute, found that United Fruit accounted for about twelve per cent of all the foreign money flowing into Guatemala, Costa Rica, Honduras, Colombia, Panama and Ecuador. 'To my mind the United Fruit Company, next to the Panama Canal, is the great phenomenon of the Caribbean world today,' wrote Willis Abbot. It was quite a comparison – it was already clear that the Panama Canal (then a year off completion) would, by linking the Atlantic and Pacific Oceans, have a profound effect on world trade.

As bananas left the region, American capital poured in. The

activities of the *fruteras* radically altered the economies of the countries in which they invested. In Honduras, bananas made up eleven per cent of the country's exports in 1892. By 1929, this figure had shot up to eighty-four per cent. Rather like the British colonial rulers in India, Americans saw this rapid expansion as a civilising force in an uncivilised part of the world.

Speaking in 1913 at the Holland House, a New York hotel, a Mr B. Gallegos, who had been living in the Colombian town of Santa Marta for ten years, talked of an 'era of progress' in the formerly deserted town. 'I am told that not until fifteen years ago did any signs appear that the place was going to awaken again but in the last decade, thanks to the efforts of the United Fruit Company, and to the English railway, the place has really progressed a great deal.' Herbert Spinden of the Peabody Museum expressed similar sentiments in 1924. Citing the ancient Mayan civilisation, Spinden told readers to 'imagine, after a lapse of many centuries, a new people filled with new hopes, with weapons against the unseen forces, entering this land and regaining its natural wealth'. The headline on his article proclaimed: 'Lowly banana rebuilds an empire.' Optimism indeed.

United Fruit established its Latin American headquarters in Panama, at Bocas del Toro ('the mouths of the bull'), an island town at the edge of a superb natural harbour on the country's Caribbean side. These days, apart from a church and a few shops and restaurants, the place is pretty quiet. The tiny airport is so close to town that arriving travellers can get to their hotel by foot, down a dusty track where ragged chickens and bright-eyed children come out to greet them. Wooden houses with carved balustrades painted in bright colours add to the town's ramshackle charm. The blue-watered lagoons and sandy beaches of Bocas are drawing more and more tourists to this somnolent corner of the Caribbean. By day, they spend their time sailing, sunbathing and snorkelling. At dusk, they gather in thatch-roofed bars by the water's edge to relax over a Balboa beer and watch the sun set.

Most of these visitors are unaware that this peaceful tropical backwater was once a thriving trading centre. The town's main avenue – where street stalls now sell postcards, shell necklaces and tie-dye fabrics – once thronged with banana workers, sailors, traders and plantation bosses. United Fruit built Latin America's first radio-telegraph station in Bocas in 1903 and three newspapers kept residents and traders informed. Smart white buildings lined the streets and five consulates were based in the town as well as a United Fruit hospital. 'From an ugly pest-hole, Bocas del Toro was transformed into one of the most healthful and attractive of tropical cities,' wrote Frederick Upham Adams in 1914.

As United Fruit's headquarters, the town was at the centre of the banana trade. More than four million bunches were exported from the Bocas plantations in 1911, according to writer Willis Abbot, with 35,000 acres under cultivation there:

> Lands that a few years ago were miasmatic swamps are now improved and planted with bananas. The great white steamships sail almost daily carrying away little except bananas. The money spent over the counters of the stores in Bocas del Toro comes from natives who have no way of getting money except by raising bananas and selling them, mostly to the United Fruit Company.

If Bocas was the nerve centre of United Fruit's rapidly expanding banana empire, its activities influenced every aspect of life in countries such as Guatemala, Honduras, Costa Rica, Panama, Colombia and Ecuador. 'El Pulpo' or 'The Octopus' owned vast tracts of land and employed tens of thousands of workers, often supplying the livelihood of whole villages and towns. The company built railways, roads, schools, medical centres and factories that produced fruit cartons. It carried out everything from mosquito control work to the securing of drinkable water supplies and the vaccination of locals. 'It is difficult to conceive of any activity – agricultural, extractive, or industrial – organized upon the basis of capital supplied from their own resources that would have yielded these countries a comparable economic return per dollar of national investment,' wrote the National Planning Association authors. With investment came influence. In Guatemala, the importance of United Fruit was such that the company ran the country's only port, Puerto Barrios, operated the Guatemalan Railroad Company as a subsidiary to its fruit business and secured a concession to administer the country's postal service.

What made all this possible was the refrigerated ship. In the late nineteenth century, refrigerated steamships had been busy. They had fed the massive demand for Argentine beef in Victorian England, where the Industrial Revolution was creating a new class of consumer. In the process, they helped enrich landowners who amassed great fortunes through the beef industry – 'as rich as an Argentine' became a phrase Europeans liked to use. The refrigerated ship also helped transform New Zealand's economy, allowing the country's lamb to become commonplace on dinner tables from London to Canada to the Middle East.

For the countries of Central America, the refrigerated steamship was also to prove a powerful agent of change. Until its appearance on the scene in the 1890s, bananas had to be stowed on the decks of slow schooners or in ventilated vessels, in which basic equipment circulated air over the fruit. This method of transport was far from ideal. Even though the fruit was picked green, it often ripened too fast. Delivering a saleable load depended on making the return journey in less than two weeks – putting seafaring traders at the mercy of wind and weather. Any delays in the passage could prove fatal to the fruit. 'Shrinkage rates', as they are now referred to in the food industry, were considerable, involving the loss of thousands of dollars worth of produce.

That was until refrigeration went mobile. The breakthrough came with the ability to produce ice or cool air with a machine. Such machines could then be installed on board vessels. Meat was the initial candidate for the new technology. In 1877, the SS *Paraguay* travelled from Argentina to France and kept its cargo of meat fresh during the voyage using an ammonia-compression refrigeration machine. Three years later, the first frozen meat shipment from Australia arrived in London and, in 1892, the first New Zealand lamb.

Soon bananas, too, were travelling in chilly holds. United Fruit introduced the first of its refrigerated vessels in 1903 and by 1928 it had amassed a fleet of eighty steamships. While meat generally travelled in holds that were kept cool by pumping cooled brine through pipes arranged in grids along the walls and ceiling, bananas were shipped in chambers in which huge fans forced air to circulate over brine-cooling batteries clustered round the base of the masts. Fresh air could be let in to help maintain the required temperature and, when approaching ports such as Boston and New York in winter, heaters were used to make sure the fruit did not suffer frost damage.

Captains found that their crews now included a new kind of maritime voyager – the engineer. Teams of experts would travel on board to ensure that the refrigeration system was working correctly. Unsurprisingly, insurance companies also started to take a keen interest. 'Should a vessel meet with a mishap, one of the first things the underwriters ascertain is whether her refrigerating plant is working,' wrote a shipping magazine in 1936. 'The temperature of a hospital patient on the danger list,' the article continued, 'is not more carefully charted than the temperature of the refrigerated cargo of a ship.' These ships were the forerunners of the high-tech reefer containers used today. They did not have at their disposal the remote temperature monitoring made possible by micro-processors and satellite communications, but they were a dramatic improvement on the method of transport Baker was forced to use when he first started sending bunches of bananas home from Jamaica.

Refrigerated transport revolutionised the banana industry. Frederick Palmer of the *New York Times*, who described the fruit as 'the most powerful American influence in Central American affairs', identified refrigerated transport as the engine driving the banana industry. In 1909, he wrote:

> The growth in consumption is only partly due to the recognition of the banana as a food. It is mainly due to improved means of transportation. The problem from the first has been to deliver the banana in edible condition at the consumer's door. Fast steamers, with their holds kept at the right temperature, now run direct to Liverpool and Hamburg.

And the right temperature is only 48°. Too much heat means that the banana will ripen too fast.

United Fruit understood the relationship between temperature and profits. It wasted little time developing its own line of vessels. Collectively, the ships were known as the Great White Fleet, named after the colour they were painted so as to reflect the hot tropical sun and help keep the bananas cool (Great White Fleet was also the term used to describe the US Navy battleships sent around the world by President Theodore Roosevelt from 1907 to demonstrate America's growing military capability). As well as temperature control, speed – delivered by steam technology – was the other prerequisite for success in the banana industry. Every hour counted. Plantation managers would be notified of the time of arrival of a refrigerated ship and the fruit would be cut – green and hard – just before the vessel docked. By the 1930s, vessels were travelling even faster, powered by a turbo-electric drive. United Fruit continually added vessels to its collection as the business grew. The company's banana boats eventually became the world's largest private fleet – and those vessels were certainly kept busy. In 1955, the fleet completed more than 5 million nautical miles, the equivalent of sailing round the world 230 times.

The gleaming white steamships also carried passengers. United Fruit expanded into the tourist industry, putting together luxury package cruises to destinations such as Guatemala, Cuba, Jamaica, Colombia and Costa Rica. It built grand hotels in the most popular destinations. In Jamaica, the Titchfield Hotel at Port Antonio, built by Baker in the 1900s, and the Myrtle Bank Hotel in Kingston, both lavish establishments, constituted the most chic accommodation available at the time. However, despite the company's successful hospitality business, its most important cargo remained the bananas. United Fruit's unofficial mantra was at one point said to be: 'Bananas guests; passengers pests.'

Fortunately for the marketing executives working to promote sales back in the USA, few consumers thought much about how their bananas got to the shops. If they had, they might have questioned some of the words of the famous 1945 United Fruit banana song, sung to a Latin beat:

> I'm Chiquita banana and I've come to say
> Bananas have to ripen in a certain way;
> When they are fleck'd with brown and have a golden hue
> Bananas taste the best and are best for you;
> You can put them in a salad,
> You can put them in a pie-aye;
> Any way you want to eat them,
> It's impossible to beat them;
> But bananas like the climate of the very, very tropical
>   equator;
> So you should never put bananas in the refrigerator.

What most people hearing this song probably failed to realise was that bananas were quite suitable for refrigeration (in the fridge, the skins darken but the fruit inside remains good to eat) – after all, they had travelled in a refrigerated ship before reaching the USA.

The banana boats were supported on land by rail and canal links. A frenzy of infrastructure building accompanied the expansion of the trade. Writing in 1941, Charles Morrow Wilson described the scene:

Jungle-busting railroad men, usually led by rough hard-cussing road masters from the United States, then charge into the wilderness with formidable assistance of 'dulldozers' or caterpillar-mounted 'draglines,' which slosh and roar as they upset trees, straddle giant logs and wallow through seas of mud to scoop drainage canals and lay railroad grades . . . Rails laid one day frequently carry trains the next.

Trains allowed the banana companies to stretch their tentacles further inland, and they hauled bananas along the tracks for increasingly impressive distances. The longest route was in western Guatemala. The fruit there rattled along railroads across the country's central mountains to Puerto Barrios, a teeming banana port on the Caribbean side of the country. In this mosquito-ridden town, the yellow cargo would pass through wooden sheds where banana-handling machinery and men worked round the clock. It then continued its journey to America or Europe in one of the ships lined up along the quayside.

United Fruit and its competitors took charge of everything along their global banana supply chains, planting, cultivating, picking, transporting, marketing and selling bananas. They were among the world's first examples of what is called 'vertical integration', in which different aspects of a business – from sourcing raw materials to transportation and marketing – are brought together. The fruit companies owned everything from steamships and railroads to telegraph lines. United Fruit was quick to see the value of being in control of all parts of the business. Indeed, what had attracted Minor Keith to the idea of partnership with Andrew Preston and the Boston Fruit Company was its ownership of the Fruit Dispatch Company, a US fruit marketing business and distribution agency. However, transportation remained at the heart of business success. A quick glance at the names of some of the fruit companies – Atlantic Fruit & Steamship, Tropical Trading & Transport Company and Standard Fruit & Steamship Company – is evidence of what drove the banana trade.

Without the temperature-controlled steamships, the business would have ground to a halt, something that became abundantly clear during the Second World War, when the US government started requisitioning commercial vessels. The banana industry was virtually shut down. 'Housewives who have been grumbling to storekeepers because of the higher prices for bananas can blame the war,' wrote the *New York Times* in 1942. '"Lack of ship bottoms," was the laconic explanation of one wholesaler, who said he is having so much trouble in getting his shipments from Central America delivered in New York that he is seriously considering going out of the banana business.'

By transporting bananas, refrigerated ships also helped shape the political landscape of Central America. The buying power of the fruit companies gave them close connections to the leaders of the Latin states in which they were investing – and with connections came privileges. For a start, they were allowed to buy enormous amounts of land. In Honduras, five foreign banana companies held more than a million acres of land on the coast. In Guatemala, large chunks of land were handed to United Fruit by the country's military dictator Jorge Ubico and wherever it operated, the company could set prices and conditions for workers. By the 1930s, United Fruit owned plantations across Central America that collectively amounted to land the size of Switzerland.

Banana companies were often given exemption from taxes such as import duties and officials would relax certain regulations in their favour. The malleability of Latin governments is understandable. After all, the infrastructure and jobs created by the banana industry had kick-started the development of many of the banana republics. Often they had not been doing much with the fruit companies' land in any case. Politicians were only too happy to sit back and watch foreign enterprises develop it. Moreover, banana companies regularly bankrolled these impoverished administrations. In 1932, United Fruit lent the Colombian government $500,000 – a hefty sum at the time.

The price that governments paid for such concessions was high, however. The loss of import duties was substantial, and while the fruit companies agreed to build railways, they frequently only constructed lines in areas that served their own interests. At the same time, politics and commerce were becoming entangled. Companies had a far greater impact on the lives of the population than state governments could ever hope to. On the north coast of Honduras, American *fruteras* were running a sort of mini-economy within the country. Corrupt officials were happy to take bribes in order to get approval for banana industry projects. It was in Honduras that 'Banana Man' Sam Zemurray would reportedly declare: 'A mule costs more than a deputy.'

Fruit companies were naturally reluctant to relinquish the privileges that accompanied their economic might. So when that might was threatened, they took action. In Honduras, a threat materialised in 1907, when Honduran president Manuel Bonilla – a great friend of Zemurray – was overthrown and replaced by Miguel Dávila. The new administration quickly entered negotiations with the USA that would allow America to take on Honduras's massive national debt, but would give the USA control of the appointment of customs officials and setting import tariffs. For Zemurray and the Cuyamel Fruit Company, this was bad news. Higher taxes on imported goods would make his railway building projects – which required materials brought in from overseas – far more expensive to execute. Zemurray decided to do something. In 1911, he financed an operation by Bonilla to overthrow the new president. When the USA (which felt no special loyalty towards Dávila) did little to intervene, Bonilla was restored to power. He and his successor, Francisco Bertrand, soon resumed the tradition of giving generous concessions to the American fruit companies.

For US consumers, the benefits of the banana boats were clear – a new cheap and plentiful source of food. The same could not always be said for the countries from which the vessels set out on the journey home. It was in Guatemala, however, that the most celebrated example of US meddling was played out – and the well-travelled banana was at the heart of the action. Having sold his Cuyamel Fruit Company to United Fruit in return for enough stock to make him the company's largest shareholder, Zemurray used his holding to start throwing his weight around and, in 1938, he became company president. By then, United Fruit was a gargantuan corporation with much of the Guatemalan economy in its clutches. Having been at liberty to do more or less what it pleased there, the company became uneasy when in 1945 Juan José Arévalo Bermejo, a schoolteacher, became president after the 'October Revolution' had ousted the dictator Ubico a year earlier. Arévalo was displaying a worrying concern for the plight of Guatemalan workers, who had at the same time started to express their dissatisfaction at United Fruit's wages and working conditions through a series of labour strikes. But if this was unsettling to El Pulpo, a far bigger threat to its commercial freedom was to come once Arbenz took over from Arévalo in 1951.

Alarmingly for United Fruit officials, the new president seemed determined to shake up the status quo. First, he started introducing elements of competition to the market. Crucially, many of the new developments involved creating alternative transportation systems to those controlled by United Fruit. Arbenz's projects included the construction of a highway to the Atlantic, where a new port at Santo Tomás would also be built. This upset United Fruit's virtual monopoly on shipments, through Puerto Barrios – and the company profited not only from transporting its own bananas. It also made money from shipping other people's goods on its Great White Fleet. Arbenz also started introducing reforms designed to improve working conditions. In the many disputes between banana workers and their bosses, the country's courts now ruled more frequently in favour of labourers.

For United Fruit, the most worrying development came in 1952 with something known as Decree 900. Part of Arbenz's land reform programme, it specified that more than 200,000 acres of United Fruit land was to be expropriated and given to peasants. The government would compensate the company with cash, but was offering an amount equal to what United Fruit had for years been claiming the land was worth for tax purposes. The company, of course, had traditionally under-valued its land. Even though much of this land was not even being used for growing bananas, United Fruit immediately protested, arguing that it was worth more than twenty times the amount being offered in compensation. At the request of United Fruit, the US government even became involved, lodging a formal complaint on the company's behalf and demanding appropriate payment from the Guatemalan authorities.

While United Fruit's frustration with Arbenz's reformist policies mounted, back home in America anti-communist sentiment was rising as McCarthyism took hold. Soon there was talk of 'reds under the bed' beyond the country's borders. The view that communism was about to sweep through Central America was not one that United Fruit discouraged. With the help of Edward Bernays, a charismatic but controversial public relations executive who was hired as the company's adviser, United Fruit had made efforts to persuade the US government that Guatemala's pro-labour stance was a clear sign of the country's communist leanings. In the early 1950s, Bernays arranged for groups of journalists from news organisations such as the *New York Times*, *Time* and *Newsweek* to visit Guatemala. Bernays believed in 'organising chaos' when it came to getting a message across. 'The conscious and intelligent manipulation of the organized habits and opinions of the masses is an important element in democratic society,' he wrote in 1928. 'Those who manipulate this unseen mechanism of society constitute an invisible government which is the true ruling power of our country.'

Back in the USA, journalists were voicing concern about the possibility that communists were operating in America's backyard. 'Guatemala today offers another unhappy example of communism sowing the wind and reaping the whirlwind,' wrote the *New York Times* under the headline: 'Communism and bananas'. Another *Times* writer, Flora Lewis, described her impressions of Guatemala City. 'Its low stucco buildings, made squat for earthquake protection, and its neatly laid streets mark it clearly as a place for gentle snoozing in the sun,' she wrote. 'Nothing roars and nothing rushes. It doesn't look or sound like the one place in the Americas where devoted, angry-tongued Communists have deeply entrenched themselves. Nevertheless, it is.' It was stirring stuff.

Meanwhile, the US administration was also keeping a watchful eye on Guatemala. The Cold War was under way and, by the early 1950s, preventing the global expansion of communism was a prominent item on the US foreign policy agenda. After all, it had only been a few years since in 1948 Stalin had blockaded West Berlin in his attempt to expand communism throughout Europe. Now, the worry was that the Soviet Union was looking around for fertile fields outside Europe in which to sow the seeds of Marxist-style revolution. Central America, however, was to be an entirely different battlefield from Berlin, where free citizens surrounded by Soviet forces were kept from communism's clutches by food supplies. The role of moveable bananas – and the companies that moved them – was to prove far more complex in this particular Cold War skirmish.

Guatemala, reckoned the US State Department, might well be a candidate for the Soviet Union's expansion plans and, as Arbenz continued to pursue social reform, its hunch grew stronger. The Central Intelligence Agency, too, viewed Guatemala with suspicion. 'The CIA saw Guatemala as a threat sufficient to warrant action,' wrote historian Nick Cullather, who had access to agency documents. 'In early 1952, analysts found that increasing Communist influence made the Arbenz government "a potential threat to US security."'

The agency did not have much to go on. It looked in vain for signs of links between Guatemala and Moscow in the form of payments or travel records. Ironically enough, the only evidence of dealings between the Soviet administration and the Arbenz government emerged when the Russians tried to buy some bananas. The sale had fallen through since the only way of transporting the fruit would have been by using United Fruit vessels. What is more, the communist presence in the Guatemalan administration was fairly weak – communists held only four out of the sixty-one seats in Guatemala's congress. Never mind that the party never signed up more than 4,000 members in a country whose population was about three million, the CIA was convinced Arbenz's land reform programme was the prelude to a full-blown embrace of communist principles.

In 1954, Milton Bracker described Guatemala in a *New York Times* article as 'a bustling outpost of Soviet propaganda right in the heart of the Americas'. But while the extent of the communist threat in Guatemala is now debated, Bracker recognised what was going on. The real danger, he wrote in the same article, was that 'the growing feeling of impatience in the United States, based on an often distorted view of what the Communists are doing here, may result in severe measures against this country'.

Three months after that article was published, 'severe measures' were indeed taken against the country. The so-called 'Liberation' of Guatemala began on the morning of 18 June 1954. It was led by Carlos Castillo Armas, an exiled Guatemalan who crossed the border with a small band of rebels in tow. 'Men who wear sun-tan trousers and shirts are flying in chartered Honduran civil transports to fields near a point where Honduras, Guatemala and El Salvador meet,' wrote the *New*

York Times. Meanwhile, planes circling the capital city dropped leaflets demanding the resignation of Arbenz. The country's airwaves were jammed and rumours that the government was losing control were broadcast across the country.

The invasion, the radio broadcasts and the leaflets had, of course, been instigated by the CIA, who codenamed the whole thing Operation PBSuccess and selected Armas as the man for the job. In the end, it was all over quickly. On the evening of 27 June, Arbenz announced that he would resign in favour of a military junta. He had been in power for four years. The army chiefs taking over agreed that their first act would be to get rid of the communists.

Arbenz was driven from power less than four decades after Captain Lorenzo Dow Baker established the Boston Fruit Company, but by then his world had radically altered. Refrigerated ships had helped the *fruteras* transport millions of bananas to American consumers, transforming Arbenz's country and others in the region, first through the development of infrastructure and the creation of jobs and, later, through political upheaval.

Chiquita, the modern-day incarnation of the United Fruit empire, remained a controversial organisation for many years. Since 1992, with the help of the Rainforest Alliance, the New York-based environmental group, Chiquita has undergone a dramatic ethical turnaround. In recent years it has been applauded – most remarkably by Ron Oswald, the general secretary of the International Union of Food Workers – for its efforts to source bananas from sustainably managed plantations where workers are paid fair wages. However, memories of labour strife and what Chiquita calls its 'use of improper government influence' linger.

The Guatemala coup of 1954 has generated forests of books analysing what happened and why. Their authors debate whether it was the commercial might of United Fruit that persuaded the US government to use communism as an excuse for regime change, or whether the USA simply used El Pulpo's grievances with Arbenz as an excuse to stamp out communism in Central America. United Fruit was certainly in a position of influence. The company had powerful connections with senior government figures such as Allen Dulles, head of the CIA, and his brother John Foster Dulles, US Secretary of State, both of whom were United Fruit shareholders. And yet after analysing the CIA records, Nick Cullather plays down the role of the company in the events of 1954. Others even say that the USA would have moved to topple Arbenz regardless of United Fruit's interests. The extent of the communist threat is controversial too – was Arbenz in the clutches of a small but influential bunch of communists, or did he simply use the communists to help instigate land reforms that were motivated by his desire to improve the lot of his people?

Whoever is right, it remains true that without the well-travelled banana and the economic clout it brought American fruit companies, Guatemala might never have come to the attention of the US administration. Certainly, it was the move by Arbenz to seize United Fruit lands and his clear preference for the interests of workers over those of banana companies that helped convince America the 'Red Peril' was on the march in Guatemala. Moreover, without the banana shipments, the giant company against which the workers felt the need to protest might never have existed. The banana changed the countries of Central America for ever, and none more dramatically than Guatemala. 'The Guatemalan economy has been linked to the United States almost as if it were a state of the Union,' wrote the *New York Times* at the time of the coup.

If Arbenz was the most visible casualty of the 1954 invasion, the real victims were the people of Guatemala. They had lost one of the first Central American presidents to put the interests of ordinary citizens above those of foreign businesses or wealthy elites. The coup, moreover, triggered a civil war that would plague the country for nearly four decades, during which death squads, kidnappings, beatings, assassinations and torture were commonplace. At first students, professionals and opponents of the regime were targeted. Later, whole villages were wiped out. An estimated 200,000 people died during the conflict. A culture of fear swept through the country that has left it permanently scarred.

Just five years before the 1954 coup, the Berlin Airlift had used moveable food to help bring about a shift in Europe's political landscape, securing West Berlin's freedom from Soviet control and thwarting the further spread of communism through Europe. In Guatemala, too, moveable food – this time, millions of yellow tropical fruits stashed in the chilly holds of refrigerated ships – was bound up with efforts to halt what the USA believed was the imminent threat of communism's expansion. However, in Central America any talk of freedom and democracy was coloured by the immense profits of the US fruit companies, and this time the people who had been 'saved' from communism had precious little to celebrate.

7

Whey to Go

Mongolian nomads practise mobile biochemistry

Yoghurt: Semi-solid food prepared from milk curdled by bacteria

Origin: The Middle East

Etymology: Mispronunciation of Turkish *yogart, yogurt,* from the verb *yogur,* to knead

Legends: Many claim to have created the first yoghurt, but one tale has it that when goats were first domesticated in Mesopotamia in about 5000 BC, the milk taken from the animals was stored in gourds. As the hot sun warmed the gourds, the milk soured, forming a curd that was found to be delicious

ERMENTED MARE'S MILK might not sound like the most appetising of drinks but to the Mongolians, *airag* is the nectar of the gods. Before setting off for battle, the great thirteenth-century warrior Kublai Khan would pour out a libation of this alcoholic liquid and raise a toast to his deities to ask for their assistance in defeating his enemies. In fact, the drink is not that bad – and in any case, the sour, milky taste and slight fizz is more than made up for by the warmth and good humour of the people serving it. Here inside their *ger* (the circular tent known by others as a yurt) a huge wooden bowl of mare's milk sits fermenting by the stove at the centre of a cosy interior. Two small children scramble around playing with a rubber ball while their mother, dressed in a bright blue *del* (a traditional coat-like tunic), dodges them as she completes the household chores.

All about the place is a cocktail of garish colour – a curious mixture of craftwork and kitsch. Oriental rugs cover the floor and brightly painted wooden furniture lines the sides of the *ger*. On a yellow ochre chest decorated with elaborate patterns is a frame containing photographs of the family, a Chinese thermos flask and a TV set. Simple iron beds are adorned with bedspreads in bright pink and red floral patterns and near the door is a clock, whose gold plastic hands move round a colour photograph of Swiss mountains. At the centre of the *ger*, a blackened stove is the point of focus. Curing in the smoke above it is a collection of objects resembling bones more closely than meat. As morsels of diced grey mutton are passed round (they look suspiciously as if they were selected from those gruesome bones above the stove), the mare's milk suddenly seems relatively appealing. But whether it is the *airag* or the snacks being dished out, the smell inside the *ger* is decidedly meaty and slightly rancid.

Welcome to Mongolia. Here in this nation sandwiched between China and Russia, the odour of mutton is hard to escape. Even in the dingy hotels of the capital, Ulaan Baatar, a hint of mutton fat lingers among the towels and bed linen. Next to the state department store, Little Hong Kong, the city's first Chinese restaurant, has a menu whose improbable dishes include sweet and sour mutton, mutton with black bean sauce and stir fried mutton with bamboo shoots. Alive or dead, sheep are everywhere in this vast, landlocked country. They are the most numerous among the country's animals, with about thirteen million of the woolly creatures grazing out on the open plains. Among them are dozens of indigenous breeds, from the stout Roman-nosed Kazakh to the Torguud, a large leggy beast with broad chest and a high-set body.

Mongolians love their mutton. They eat it semi-raw or boiled and unsalted, preferably with a thick wedge of fat still attached to it. Mongolians eat more mutton and goat than any other nation in the world, beating the other great lovers of these meats, the Australians and New Zealanders. In 1999, each Mongolian ate an average of more than 100 pounds of mutton and goat meat, compared with about 64 in New Zealand and 35 in Australia. The link between Mongolians and mutton is so strong that when the Chinese developed a winter dish now popular in Beijing restaurants – slices of meat are placed in a copper pot containing a boiling soup base – they called it 'Mongolian hotpot', even though the recipe is unknown in Mongolia.

As well as consuming meat in large quantities, Mongolians

also foster variety. Sheep, camels, cattle (including yaks), goats and horses are all raised on the steppe. Moreover, it is not just the flesh of these beasts that provides sustenance for the country's nomads. Dairy products make up a substantial part of a diet that, at times, does not even include meat. Because they eat such a diverse range of dairy foods – from cheese and curds to yoghurt and mare's milk – from such a range of different animals, these foods are sufficiently full of nutrients and vitamins to make the addition of vegetables unnecessary. Vegetables, if consumed at all, are served in tiny amounts. Vegetables, Mongolians will tell you with an air of disdain, are for the weaker, sedentary peoples of the world.

Unlike consumers in Europe or America, Mongolian nomads do not buy food that has been sterilised and put in cartons or retort pouches that give it a shelf life of several weeks. Meat is not cut into neat pieces, given artificial colouring and vacuum-packed. Yoghurt does not come in small plastic pots with a picture of a strawberry on the side. Nor would the kind of yoghurt found in a supermarket be sufficiently nutritious to become the key component in a diet, despite marketers' declarations of how many probiotic ingredients or Omega-3 fatty acids it contains.

Yet the yoghurt eaten by Mongolians and the one we pick up in the chilled cabinet of the supermarket do have one thing in common: they are both extremely well travelled. As interest mounts in the concept of 'food miles' – which questions the reliance on long-distance food transportation and, because of the carbon emissions associated with it, is at the forefront of climate change debates in countries such as Britain – several studies have focused on yoghurt and the number of miles travelled by each of its components, from the flavourings to the plastic pot and foil lid. However, while transport of each of the elements in mass-produced yoghurt involves burning fossil fuels, the travels of Mongolian yoghurts and other dairy products rely on a completely different system: live animals. After all, if your food has legs, you do not need to invent elaborate mechanisms to move it about – you have food that moves itself.

For centuries nomadic Mongolian families have travelled around with the source of their food: animals. Known as pastoralism or transhumance, the idea is to move with your livestock to fresh pastures as the seasons change. It is a system that has provided an astonishingly efficient means of survival in a country with some of the planet's harshest conditions and most dramatic swings of climate. In this land of extremes, it seems improbable that humans can survive at all. In the Gobi desert, summer highs of 50° Celsius plummet in winter to –40. During particularly brutal winters – known as the *dzud* or 'white death' – ice crusts and snow cover the pastures causing thousands of animals to die. Sometimes, the creatures expire while standing, frozen solid by the bitter Arctic winds. This is the environment in which Mongolians have lived successfully for centuries.

Modern infrastructure is scarce in Mongolia. In a country the size of France and Spain combined, there are only a thousand or so miles of paved roads and fewer than six fixed-line telephones per hundred citizens (though mobile phones are becoming popular). What Mongolia does have, however, is grass – and lots of it. Out on the steppe (a semi-arid grass plain devoid of trees) it is possible to travel for days without seeing a building or even a power line. In the summer, flowers for a brief and wonderful moment transform the landscape. Almost overnight, edelweiss and columbine cover the steppe with delicate mists in vibrant hues. But this explosion of colour is merely a short break from the waving sea of green that dominates the view for much of the year. Grass is everywhere you look in Mongolia. It rises up with a vigour that seems to defy the harsh environment in which it grows. Grass sucks in the energy of the fearsome sun that beats down on the 'Land of Blue Sky'. It is what powers life on the steppe.

This is how the system works: as the grass matures, its nutrients, acquired from the sun by means of photosynthesis, nourish and fatten the animals that Mongolians then use as their main source of food. Indigestible to humans (we lack sufficient numbers of stomach chambers), grasses and their nutrients are passed on to the Mongolians via their animals – and they have plenty of animals. In Mongolia, animals out-number people by about ten to one. Here, wealth is measured not by your house, land or possessions but by the size of your herd. When the seasons change, you simply move your bestial riches to new pastures to find fresh fodder. Because the animals are not constantly grazing on the same patch, the grasslands can recover in time for the next feed.

It is a pattern of life that has changed little since the thirteenth century, when Marco Polo observed Mongolian nomads on his travels:

> They spend the winter in steppes and warm regions where
> there is good grazing and pasturage for their beasts. In
> summer they live in cool regions, among mountains and
> valleys, where they find water and woodland as well as
> pasturage. A further advantage is that in cooler regions there

are no horse-flies or gad-flies or similar pests to annoy them and their beasts. They spend two or three months climbing steadily and grazing as they go, because if they confined their grazing to one spot there would not be grass enough for the multitude of their flocks.

Moving home every few months is made less troublesome by the *ger*. This cylindrical residence can be erected or dismantled in a matter of hours and packs neatly into wagons and trucks. It is an ingenious arrangement. Walls consist of a collapsible latticework wooden frame. Above them, a mass of red spokes converges like the underside of a giant umbrella to create the domed roof. At its centre a circular opening allows smoke from the stove to escape. A wooden door secures the entrance, often decorated with vibrant colours and inventive patterns. The whole thing is encased in a felt membrane of sheep's wool with a heavy white canvas cover on top. The *ger* is engineered to cope with Mongolia's extreme temperatures. Its dome shape deflects the bitter winds that sweep down from Siberia in winter while in summer the felt and canvas cover can be rolled up from the base to allow air to circulate through the interior. Marco Polo was certainly impressed. Mongolians, he explained, 'carry [the *gers*] about with them on four-wheeled wagons wherever they go. For the framework of rods is so neatly and skilfully constructed that it is light to carry.'

Because of their transitory lifestyle, Mongolians have few possessions compared to settled peoples. A few pieces of furniture, a variety of cooking pots, perhaps a television set, and that is about it. However, any notion that nomadism is a primitive way of life is put to rest when you start to examine the complexities of mobile farming. Unlike traditional

European herding, where flocks are simply moved from high summer pastures on alpine slopes to low ground in winter, Mongolian conditions present herders with all sorts of tricky challenges. In a country where the wind-chill factor has to be reckoned with, shelter is a key factor in the choice of pasture, particularly since livestock are not housed. Mongolians also have to find appropriate altitudes for their livestock. While yaks are resourceful creatures that can graze through thick carpets of snow and so thrive at higher altitudes, cattle flounder at the first sprinkling of snowflakes.

Certain animals prefer certain types of grass (among the choices are needlegrass, tatary buckwheat, wild oats and squirrel-tail barley), and the complexity does not end there. Lactating animals need more grass than those without young offspring. Different animals require grass that is at different stages of growth. Sheep are the least fussy about what they eat. However, even sheep require careful control for if they are left too long on a particular piece of pasture, they can ruin it, since they tend to concentrate on the grasses they like best, leaving the poorer varieties to go to seed and spread. Considering the additional challenges of dramatic daily and seasonal temperature swings and an astonishingly diverse topography, from mountains to grasslands and deserts, the nomadic life starts to look far from simple.

Another myth is that nomads wander at random across an unlimited area in search of greener pastures. In fact, the traditional Mongolian system was highly organised. For a start, Mongolians believe that their ancestral spirits own specific areas of land, and staying within their own four-season pastureland gives people a strong sense of a homeland – even though that homeland may stretch across several hundred miles. When Mongolians leave their region, tradition stipulates that they carry stones from it to honour the local gods of their new home. Nostalgia for the homeland is a frequent theme in the ballads of the country's celebrated throat singers – musicians who, through a mysterious technique involving muscular control of the abdomen, chest and throat, produce two voices simultaneously.

Nor is access to land ungoverned. Traditionally, political authorities regulated the pastures, with the ruling families of what was a sort of feudal system controlling the allocation of land. 'Every captain, according to whether he has more or fewer men under him, knows the limit of his pastorage,' wrote William of Rubruck, a Franciscan monk who visited in the thirteenth century, 'and where to feed his flock in winter, summer, spring and autumn.' It was a neat solution to sharing land needed to sustain life and it continued largely unchanged until the 1920s.

That was when a giant bulldozer barged its way into this efficient moveable lifestyle: Soviet-style communism. In 1921, Mongolia won independence from China, but freedom from Han domination had strings attached. It was the Russian Red Army that assisted the Mongolian independence fighters and the Soviet influence extended well beyond the battlefield. When the Mongolian People's Republic was founded in 1924 it was as the world's second communist state. With China a constant threat from below, Mongolia accepted the Soviet embrace as a means of maintaining its independence.

Such a close relationship meant that Mongolia followed the social and economic path of its powerful northern neighbour – and it was a heavy debt to pay. In the late 1920s, socialism swept rapidly across the country with devastating conse- quences. In a disastrous collective farm experiment (which was ultimately abandoned), 600 feudal nobles and monasteries had their wealth – that is, their herds – handed over to the collectives. Then in the 1930s, thousands of monks were killed in Stalinist-style purges, many being forced to dig their own graves before being buried in them. Most of the country's Buddhist temples were smashed to the ground or left in ruins. Mongolia remained a puppet state until the Soviet empire's collapse in 1991. Meanwhile, a strange thing happened. Although forced collectivisation had been abandoned, a voluntary system encouraged nomads to place their herds within pastoral collectives. This gave them access to a pool of funds that could be used to build winter shelters for animals, dig wells, provide veterinary services and buy supplementary fodder. The initiative proved popular. By the late 1950s, most Mongolian herders were part of a collective system.

Communism in Mongolia was a brutal experience. But the country's odd, hybrid version of collectivism created a phenomenon not seen elsewhere. Because the traditional nomadic system regulated the use of land, the idea of sharing land – a concept at the heart of collectivisation – was familiar to Mongolians. Land was something they used at different times in different seasons. Property consisted of livestock, not terrain. In fact, it was when Soviet communism collapsed in the 1990s and a free market economy was introduced to Mongolia (with the assistance of a bunch of Harvard graduates) that the trouble started. Elsewhere, as communism imploded, former Soviet citizens set about reclaiming the property that had been seized from them by the state. In Mongolia, however, attempts to privatise land met opposition. Moreover, once the Soviet system disintegrated, herders lost the services provided by the collectives. It became necessary to squeeze more income from the pastures. In the 1990s, as the more commercially driven Mongolians increased the size of their herds, over-grazing became common and the ancient system of land sharing started to fall apart.

For Mongolians, this system is more than just a means of providing food in a hostile environment. Their cultural identity is rooted in the nomadic lifestyle. The habit of living in tents is hard to abandon, even in the city. In Ulaan Baatar, *ger* suburbs are among the more striking sights in this eccentric capital. The round tents cluster together in large groups on the outskirts of town, often hooked up to electricity and water supplies. Anthropologist Uradyn Bulag argues that the pastoral lifestyle is so central to the culture that 'anybody who is not raised in such milieux is seen as non-Mongol, and even as biologically something different'. Nomadism has also distinguished the Mongols from their more sedentary southern neighbours, the Chinese – something that may even explain the Mongolian disdain for pork, since pigs need settled conditions to thrive.

The idea of living with your food has led to some curious culinary habits. In what was the ultimate in portable rations, Genghis Khan, Mongolia's great thirteenth-century conqueror, advised his soldiers that when food supplies ran short, they should stick a straw into the neck of their horses and drink the blood. A soldier could survive for several days on this diet. The ancient Mongolian military is not alone in this practice. Masai, the East African warrior tribe of nomadic pastoralists, are perhaps most famous for drinking the blood of their cattle. In Elspeth Huxley's 1959 childhood memoir, *The Flame Trees of Thika*, she describes how Sammy, a Masai foreman, would bleed his animals:

A kikuyu seized the head of a brindled bull and twisted it over his thigh, gripping its neck with one hand so as to swell the jugular vein. Sammy took a bow from the boy's hand and, from a few yards' range, fired an arrow straight into the jugular. The arrowhead was ringed with a little block of wood so that its point could not penetrate more than about half an inch. Still with a casual air, Sammy plucked out the arrow and the blood spurted into a calabash held by the boy. Then Sammy closed the arrow-prick with finger and thumb and, to my surprise, it stayed closed and the bleeding stopped. The bull, released, strolled off and started to graze.

If this technique might sound a little unsavoury as a means of obtaining a snack, what was done with the blood sounds even more disagreeable. 'Sammy did not drink the blood in the calabash. The boy mixed it with milk and other ingredients, of which cows' urine was one, and let the brew ferment for a day or two. When it was ready to eat, its consistency was like that of soft cheese.'

The dairy element of this unusual recipe is another thing that the Masai have in common with the Mongolians. Dairy products are at the heart of the Mongolian diet and, at times, it is all they eat. It is also an important element of social life. At festivals, a generous libation of *airag* is a crucial part of the proceedings. Milk products are deemed to have spiritual properties and have become part of religious rituals. In the morning, milk is sprinkled around the *ger* nine times using a long wooden spoon. This ritual must also be performed before embarking on a journey, with the ceremony conducted by the departing traveller while facing their destination.

Dairy products are easy to store and convenient to carry, something that did not go unnoticed by Genghis Khan and his warriors. When they were not drinking horse's blood, soldiers would carry with them a solid paste produced by bringing milk to the boil and skimming off the cream (this was saved for making butter). Then the milk would be left in the sun to dry. 'When they are going on an expedition, they take about ten pounds of this milk,' Marco Polo explains, 'and every morning they take out about half a pound of it and put it in a small leather flask, shaped like a gourd, with as much water as they please. Then, while they ride, the milk in the flask dissolves into a fluid, which they drink.' Churned up by the movement of the cantering horses, the drink was what must be the world's earliest milkshake.

Surprisingly, the one thing Mongolians tend not to consume is milk itself, fresh from the udder. Instead, these people have become masters of biochemistry. Carefully controlling the action of microbes, they produce an astonishing range of fermented dairy products that they can carry around with them and that remain edible for long periods of time. In his essay on Mongolian dairy products, Rinchingiin Indra describes the various foods produced: yoghurt, cheese, butter and fermented mare's milk, *ezghee* (made from milk curds), *aragoul* (a dried curd product made in the autumn, when the animals' milk supply is thickening and drying up) and *shar tos* (butter taken from melted cream skins). An alcoholic drink called *nermel* is made from fermented milk. These dairy products do not just add variety to the diet. They are also high-calorie foods packed with nutrition.

It is the extraordinarily varied sources of milk that make the Mongolian nomadic diet so nutritious. The Mongolian

whey to go

nomads' mobile dairy contains a diverse collection of animals. Cows and sheep are not the only creatures the nomads are prepared to milk. They also tap into the udders of horses, camels, goats and yaks. There is even a sort of hybrid animal – a mixture of a cow and a yak – that is raised in Mongolia and, of course, is also milked. It is often said that the Chinese taste for exotic animals is such that, if a creature's backbone points to the sun, they will eat it. In the case of the Mongolians, you might say that, if the animal has an udder, they will milk it. Each animal yields liquid of a slightly different flavour and with different nutritional properties. Yak's milk, for example, is prized for its high fat content, and it can be turned into protein-rich products such as cottage cheese and yoghurt. Because yak milk contains bigger fat globules than those in the milk of other animals (almost double that of Swiss Simmental cows, for example), less effort is needed to turn it into butter.

However, protein and fat content is not always the priority. Mare's milk is lower in protein and fat than milk of other animals, so when it is fermented to create *airag* it produces a liquid that is easy to digest – so much so that some Mongolians find they can drink more than twenty pints of it a day. William of Rubruck described this ubiquitous Mongolian tipple. 'It is pungent on the tongue like rapé wine when drunk, and when a man has finished drinking, it leaves a taste of milk of almonds on the tongue,' he wrote. 'It makes the inner man most joyful and also intoxicates weak heads, and greatly provokes urine.' In spite of the inconvenience of frequent trips to the bathroom, *airag* is extremely healthy. It contains four times the number of fatty acids in vegetable oil and lard and the fermentation process increases the milk's already rich vitamin content. Mare's milk is said to improve the complexion and eyesight.

What is more, unlike other animals, horses are immune to diseases such as tuberculosis and brucellosis – a nasty flu-like infection transmitted to humans through consumption of animal products.

There is, however, a price to pay for the benefits of this nutritious liquid: because of their small udders, mares must be milked constantly. Foals feed up to seventy times a day. As the baby matures and turns to grass, the mother must be milked throughout the day – and milking a mare is a tricky business. Rinchingiin Indra describes the process:

> The milkmaid approaches the animal from its left side and gets down on her right knee. She places a milk pail on her raised left knee and stabilizes it with a string drawn tight to her left forearm. She brings her right hand to the udder from the rear and brings her left hand around the front of the mare's thigh as if her arms were embracing the animal's leg. Milking is always begun while the foal is touching the mother's side. After the flow has started, the milkmaid takes over. Another person holds the foal away from the udder, but the offspring continues to remain in physical contact with the mother until the milking is over.

If this sounds complicated, so are the biochemical processes required to turn raw milk into Mongolian dairy products. The tools Mongolians deploy – everything from cauldrons and ladles to wooden tubs and cow-skin sacks – may look primitive, but the basic chemical activity they are setting in action is no different from that found in most commercially produced yoghurts. 'Good' bacteria are at the heart of the process. During fermentation these bacteria produce lactic acid, which helps to preserve the milk (another reason fermented foods are ideal for the nomadic lifestyle). To start with, the raw milk is heated to about 45° Celsius. This sparks the development of the rod-shaped bacteria – *lactobacillus* – that live in milk's acid. The liquid is then poured into a container and sealed tightly to keep out air. The bacteria work away furiously and within about five hours will have turned the milk sour. As the milk thickens, a second bacterium – *streptococcus* – is activated, generating the fragrant aroma that makes yoghurt so appealing. Once the liquid has solidified, the vessel's cover is removed to allow it to cool.

Many believe that the harnessing of these immensely complicated microbial cultures first occurred on the steppes of Central Asia. Yet no one has so far explained how such sophisticated processes evolved in the hands of people with no access to scientific equipment or modern technology. This in itself is impressive enough. Here on the steppe, however, biotech skills are just part of a remarkable lifestyle in which the nutrition that sustains humans is obtained from the natural resources available in the most forbidding of environments. Through domesticated animals and the journeys they make during their lifetimes (chomping through rich foliage along the way) the sun's energy, stored in thousands of acres of Mongolian grassland, is passed to humans, enriched by the digestive juices of the animals themselves and further enhanced by the process of milk fermentation. It is the most fundamental of food chains.

Mongolian animals have had to travel impressive distances to produce the milk with which nomads make their yoghurt. However, the ingredients used in commercial yoghurts collectively clock up many more miles. In 1993, a German researcher decided to track some of these journeys. Stefanie Böge, then working at the Wuppertal Institute, was looking for a way to measure the environmental impact of transporting consumer products. She hit on the idea of examining the journeys made by the elements of an industrially produced tub of strawberry yoghurt before it was sold in a shop in southern Germany. The study included everything from the yoghurt cultures and basic ingredients, such as milk, jam and sugar, to the glass container, the paper label and the foil top. The boxes the yoghurts came in were also counted, as well as the glue needed to construct them.

Böge estimated that each purchase of a small tub of strawberry yoghurt generated 30 feet of movement by a fully laden truck. This does not sound like much until you consider what happens when thousands of shoppers each go out and buy a pot, or perhaps several, of that yoghurt. The most striking element of Böge's work, however, is a map of Germany that she included in a paper titled *The Well-Travelled Yoghurt Pot*. On the map, Stuttgart, the city where the yoghurt was eventually sold, is at the heart of a crazed mass of lines that run to and from the sites of the various suppliers. Alongside the lines on the map are explanations of what was transported on each route.

Strawberries, for example, came in from Poland and paper for the cardboard boxes from north-west Germany. As the dotted lines reveal, the routes of the different components going into the yoghurt were by no means all direct. Subcontractors also received their raw materials from myriad sources. For example, three lines on Böge's map – indicating the transit of the corn starch and wheat powder used to thicken the yoghurt's texture – converge on Düsseldorf, before heading south towards Stuttgart. The strawberries from Poland travelled to Aachen to be turned into jam before being transported to Stuttgart.

What makes Böge's findings surprising is the fact that the yoghurt whose journeys she chose to follow was hardly the most intensively processed – it was a simple yoghurt with a bit of jam added. Moreover, the study was conducted more than a decade ago. If someone were to follow one of the products on today's supermarket shelves, the results might be even more startling. Such is the overwhelming array of yoghurts on sale now that it would be hard to know which one to select for such a study – perhaps the one with muesli added, the soya-based version, the kefir (a yoghurt-like drink), the frozen yoghurt (which comes in hundreds of flavours) or the raspberry probiotic yoghurt. Tracking the journeys of all their ingredients would surely keep a large team of researchers fully occupied and a map like Stefanie Böge's could end up with Germany buried beneath a dense mesh of transport lines.

Even yoghurt that is processed near the place it is to be sold clocks up the miles. In 2005, twelve years after Böge's pioneering study was published, the Leopold Center for Sustainable Agriculture, Iowa, worked out that the primary ingredients for an 8-ounce container of strawberry yoghurt processed in Des Moines, Iowa, and sold in a supermarket nearby had travelled more than 2,200 miles, with the average for each ingredient nearly 280 miles – and this was not taking into account the transport for things such as the plastic cup, lid, foil and cardboard case, nor the active cultures and flavours.

Demand for yoghurt has risen sharply over the past two decades. In the USA, where consumption was an average of just a tenth of a pound a year per person in 1955, that had increased to about 4 pounds a year by 1985 and more than 7 pounds by 2002. Dozens of brands of yoghurt emerged during this time, each with a portfolio of dozens of types of yoghurt. Stonyfield Farm's versions include organic and wholemilk yoghurt for babies and toddlers, frozen yoghurt, smoothies and cultured soya, as well as flavoured yoghurts with names like Chocolate Underground, Key Lime, Apricot Mango and Strawberry Cheescake. Yoplait, owned by General Mills, has a range of yoghurts for children called Go-GURT in flavours such as Strawberry Splash and Cool Cotton Candy. They come in portable tubes that, claims the company, makes them 'the perfect snack for on-the-go kids'. Mongolian nomads might well approve.

Manufacturers choose from a vast number of ingredients with which to enhance the taste and 'mouthfeel' of their yoghurts – anything from corn syrup, dextrose and fructose to carob gum, potassium sorbate, gelatin, modified food starch, agar, citric acid and artificial flavours and colours. Putting together such ingredients are food technologists and researchers. These scientists shape an increasingly large amount of the food on supermarket shelves, and yoghurt is one of the products on which they focus much of their inventive attention. It is not food – it is chemical engineering. Some people might be surprised to find, for example, that fish oil is found in some yoghurt. Omega-3, a fatty acid that is thought to improve brain development and enhance eye function, is being used to fortify many of the yoghurts now on the market. Omega-3 is actually derived from an oil that fish produce when they eat certain marine algae. Fish oil in yoghurt? Nomads in landlocked Mongolia might find it hard to believe, but after years of research, micro-capsules have been developed to prevent the fish oil from oxidising the yoghurt and giving it a peculiar smell.

Thousands of ingredients go into yoghurts today. Some are artificial. Some are natural. Some are health giving, some not so nutritious. All of them, however, have to travel large distances to the manufacturer's facility before taking their final journey as finished products to our fridges. Fructose might have been refined from corn grown on vast fields in Iowa. Omega-3 has travelled hundreds of watery miles in a live fish before being squeezed out in oil. Carob gum comes from an evergreen tree native to the Middle East that is also grown in Greece and Italy.

Böge's investigations into the journeys of yoghurt were among the first signs of increasing concern about the impact of globe-trotting food on the environment – particularly on the emission of carbon dioxide, which is one of the main contributors to global warming. Soon, the distances being travelled by con-sumer groceries were being scrutinised. Some people started calculating the miles collected by long-haul fruits and vegetables. Others called for a return to eating locally produced food. Attention focused less on the kinds of journeys tracked by Böge – journeys arising from the processing of food – than on the long-distance movement of fresh produce carried around in refrigerated containers.

In 1994, a year after Böge published her first study, the term 'food miles' appeared in a study called *The Food Miles Report: The Dangers of Long Distance Food Transport* published by Sustain, an organisation that promotes sustainable agriculture. All kinds of statistics started emerging from this organisation and others on the mileage acquired by fresh produce on its way to the supermarket. In another report called *Eating Oil*, Sustain calculated that a sample basket of twenty-six products could collectively have travelled a distance equivalent to six times round the equator. In the USA, the Leopold Centre pointed out that the typical American prepared meal contains ingredients from at least five countries. It found that the average total distance for three meals whose ingredients were sourced and eaten in Iowa was about 1,200 miles, compared with almost 12,600 miles if those ingredients were brought in from outside the state from conventional sources. Suddenly, the incredible journeys being undertaken by food were becoming clear, highlighting the worrying fact that the global food supply relies heavily on fossil fuels.

Concern about food miles has been at its most intense in the UK. As a small island Britain is acutely aware of the sea and air transport needed to bring food to its shores. Imports make up a large proportion of what is consumed by the nation. Yet this phenomenon is not so new. In 1936, describing the benefits of advances in refrigeration, a magazine called *Shipping Wonders of the World* reported that more than half the beef and mutton consumed in the UK was being imported. Ninety per cent of the butter on British tables was foreign, it said, and seventy-five per cent of its apples came from overseas. The magazine's tone was upbeat, applauding the new food choices made possible by modern technology.

In our post-industrial world, however, this optimism has been replaced by growing concerns about global warming and food miles have come to represent much of what is wrong with the food supply. Headlines in British newspapers have reflected this view. 'The contents of the average European shopping trolley travel 2,200 miles,' wrote the *Independent* in 1993. 'A typical, traditional British Christmas dinner has travelled 43,000 miles before it reaches your plate,' reported the London *Evening Standard* in 2005. Articles such as these have none of the celebratory tone of *Shipping Wonders of the World*'s writer. Neither do their headlines: 'Round-the-world dinners damage the environment' was one. '"Food miles" eat up money and energy' was another.

As the long-haul journeys of food have come to light, the idea of eating locally produced food has gained momentum – and not just in the UK, where there are now more than 500 farmers' markets. While in the USA the food miles issue has not been so widely publicised, farmers' markets are opening across the country. Between 1994 and 2004 the number rose by 111 per cent. LocalHarvest.org has information about the thousands of sources of local produce across America. Type a zip code into the search field and the database comes up with a list of suppliers in the area. At markets such as San Francisco's well-known Ferry Plaza Farmers' Market – now a tourist attraction that draws thousands to the city – stalls packed with freshly picked seasonal goodies from nearby farms certainly look far more appealing than plastic supermarket containers and vacuum packs containing products that have been trucked in from miles away.

Locavores, as some call themselves, have taken extreme measures in their pursuit of local food. Gary Paul Nabhan, an author and naturalist, spent a year in which four out of five of his meals were made only of ingredients found within a 250-mile radius of his Arizona home. He documented his experiences in his 2002 book, *Coming Home To Eat*. In the UK, Ollie Rowe, a talented young chef, started a restaurant with the intention of sourcing all his ingredients from within reach of the London Underground.

As with most debates about the food supply, the issue has attracted some impassioned spokespeople. Colin Hines, former head of Greenpeace International's Economics Unit and author of *Localization: A Global Manifesto*, has argued that governments should promote local food production rather than global trade. In a 1996 article in the *Nation*, Hines and co-author Tim Lang, professor of food policy at London's City University, proposed a new set of import and export rules. 'These should be introduced on the national and regional bloc levels, with the aim of allowing localities and countries to produce as much of their own food, goods and services as they possibly can. Only goods that cannot be provided locally should be obtained nationally or regionally, with long-distance trade the very last resort.' In 2001, in a *Financial Times* commentary, Caroline Lucas, a Green Party member of the European Parliament, suggested that European leaders should introduce a localist rural and food policy. 'Its goal would be to keep production closer to the point of consumption,' she wrote.

Martin Wolf, chief economics commentator for the *Financial Times*, dismisses this sort of talk as nonsense. 'Hysteria must be rejected,' he declared in 2001. 'This generation of citizens of rich countries enjoys a healthier and more varied diet than any in history. Still less should one indulge in the reactionary belief that anything smacking of trade, modern technology or large-scale enterprise must be bad. Farming's future cannot possibly lie in a return to the local self-sufficiency of an impoverished past.'

Even Adam Smith, the eighteenth-century economist, weighs in on the debate. 'By means of glasses, hotbeds, and hotwalls, very good grapes can be raised in Scotland,' wrote

Smith in 1776 (as cited by Wolf in his editorial), 'and very good wine too can be made of them at about thirty times the expense for which at least equally good can be brought from foreign countries. Would it be a reasonable law to prohibit the importation of all foreign wines, merely to encourage the making of claret and burgundy in Scotland?'

Meanwhile, local food proponents continue to state their case. Some say food security is undermined by the reliance on imports, pointing out that the average city has only two or three days' supply, making it vulnerable to disruptions such as a trucker's strike or a terrorist attack on a port. Others claim eating local food supports small farmers. Members of the Slow Food movement, founded in Italy in 1986, point out that local and seasonal produce tastes better. Most noticeably, long-distance food has become a prominent part of discussions about climate change, with many advocating a return to eating what is grown and produced locally. But is this the most realistic way of saving the planet?

Not necessarily. Importing tomatoes to the UK from Spain actually consumes less energy than keeping greenhouses sufficiently hot for tomatoes to thrive during the cold, soggy British winter, according the UK's Department for Environment, Food and Rural Affairs. Similarly, food ethicists Peter Singer and Jim Mason calculate that the total energy used in growing and shipping a tonne of rice from Bangladesh to San Francisco would be lower than that of rice grown in California under irrigation. Moreover, locally grown fruit and vegetables rack up food miles of their own unless they are consumed nearby. While consumers are putting supermarket chains under pressure to offer a range of 'local' produce, this would precipitate the arrival of dozens of vehicles from different places at the store every morning rather than one large truck from a regional producer.

The contribution of an increasingly mobile food supply to energy consumption, and the resulting emission of greenhouse gases, is not to be underestimated. However, transportation is only partly responsible for food's environmental footprint, and it is not always the biggest part. One US study cited by Singer and Mason found that cooking and preparation generated twenty-six per cent of the energy used in the food chain while processing generated twenty-nine per cent. Food transportation was responsible for just eleven per cent.

Agricultural production processes are behind much of the environmental damage resulting from what we eat. Because electricity generation is the single largest source of carbon dioxide emissions, farm equipment and vehicles and milking machines all contribute to climate change, as do farming practices such as drying crops, keeping poultry in heated buildings or cultivating fresh produce in hot houses. The manufacturing of chemical fertilisers and animal feeds also consumes large amounts of energy, releasing yet more carbon dioxide into the atmosphere.

Along with the carbon dioxide emissions arising from agriculture come other greenhouse gases. Animal manure, soil management and heavy use of synthetic nitrogen fertilisers in crop production all contribute to an increase in nitrous oxide emissions, which are up to 300 times more effective at heating the atmosphere than carbon dioxide gas. Rising demand for crops such as soya and coffee means vast areas of land in countries such as Brazil, Argentina and Vietnam are being turned over to agriculture, often leading to deforestation. Cutting down forests not only deprives the earth of the trees and plants that soak up carbon dioxide. If they are burned in the process of land clearance, they return to the atmosphere as carbon dioxide.

What is more, in agriculture, greenhouse gases come from the most unlikely places. Curiously, while Mongolians have few cars, even their system of food production contributes to climate change – through cow farts and sheep burps. And a tonne of methane warms the planet more than twenty times as much as the same amount of carbon dioxide. When emissions from land use are also taken into account, livestock, which now occupy about thirty per cent of the planet's surface, generate more greenhouse gas emissions than transport, according to a 2006 United Nations study. Mongolian animal emissions are so far not significant since nomads do not engage in intensive farming. Elsewhere, however, the problem can be a serious one. Residents in the San Joaquin Valley in the centre of California (home to more than 2.5 million cattle) complain of a thick cloud of smog that hangs over the valley, making air quality unhealthy – and very smelly.

The cars in which we drive to the supermarket sometimes put more carbon into the atmosphere than the journeys of the foods we buy there. A 5-mile shopping trip in an average-sized car to buy 66 pounds of food would emit the same amount of carbon dioxide as shifting that food almost 600 miles by truck and more than 23 miles by plane, according to Sustain's *Eating Oil* report. Singer and Mason reckon that driving an extra 5 miles to visit a local farm shop generates the same amount of emissions as shipping 17 pounds of onions from New Zealand to London.

Compared to the emissions associated with production, processing, shopping by car and cooking, transporting food generates a relatively small proportion of the greenhouse gases associated with what we eat. A more effective way of assessing food's damage to the planet, argues New Zealand's Lincoln University, is to look at its entire energy use – from production to transport, storage and waste. A study the university conducted in 2006, which looked at everything from electric fences to farm sheds, tractors and animal feed, found that dairy and lamb production in New Zealand was more energy efficient than the British equivalent, even when the transport was included.

Some companies are exploring this approach as they try to reduce their 'carbon footprint'. The UK supermarket chain Tesco announced in early 2007 that it would develop a carbon footprint measure – allowing shoppers to compare products on their emissions levels in the same way as they look at price or nutritional value – and has enlisted Oxford University's Environmental Change Institute to assist in the process. Tesco will need all the help it can get. Calculating the carbon footprint of consumer products is an extraordinarily complex endeavour – particularly for food, the energy consumption of which includes everything from producing fertilisers to heating greenhouses, running food-processing plants and manufacturing plastic yoghurt pots. As has been indicated earlier, when all these factors are taken into account, long-distance transport is not the greatest part of the carbon equation. And while constant truck trips to and from processing plants might create unnecessary food miles, the journey of a vessel such as the *Emma Maersk*, stacked with thousands of large steel containers, is a relatively small part of the picture.

As global populations converge on cities, we may have little choice but to rely on transport for our food supply. With more people living in cities than outside them (2007 is expected to mark the turning point, according to the United Nations), producers and consumers of food will become more widely separated. One exception is Portland, Oregon, where planners have long waged a war on urban sprawl, leaving fertile farmland on the city's doorstep supplying locally grown organic food. Gazing across the Tokyo skyline, it is hard to see how the Portland model could be applied. The sea of concrete stretches as far as the eye can see, housing millions of families, all requiring three meals a day. In the mid-1960s, Tokyo became world's first 'metacity' – the UN's term for sprawling urban centres of more than twenty million people. By 2020, Mumbai, Delhi, Mexico City, Sao Paulo, New York, Dhaka, Jakarta and Lagos will also be metacities.

With fewer than a million people living in Ulaan Baatar, it is unlikely that Mongolians, as they chew on a tasty piece of mutton fat, have given much thought to all of this. Nevertheless, they are part of this global demographic shift. Former nomads are moving to the city. About thirty-eight per cent of Mongolia's citizens now live in Ulaan Baatar (the only other capital city to house such a large proportion of a country's population is Bangkok, where almost a quarter of Thais live). Some of Mongolia's new urbanites have chosen the less arduous lifestyle of the city dweller over the nomadic life. Others are impoverished former nomads whose herds perished in recent natural disasters. Between 1999 and 2001, Mongolia was hit by a series of *dzuds*, killing twenty-two per cent of the country's animals, leaving nearly 10,000 families with no animals at all. As nomadic herding becomes less viable as a means of supporting the population, the knowledge and skills built up over so many centuries could be lost. Then Mongolians may find that, like the rest of us, they have to get their daily supply of yoghurt from a well-travelled plastic pot.

So if it is not possible to halt all the journeys of our food-stuffs, how can we move them about in a way that does less damage to the planet? As engineers, business chiefs and politicians ponder this question, biofuels have emerged as alternatives to gasoline – and curiously enough some of them are byproducts of foodstuffs. Chip fat can now be turned into biodiesel, for example, while another option, ethanol, is made from plant matter such as corn, soya beans and sugarcane – material that is grown (unlike oil, which is extracted from the earth's diminishing supply). While ethanol generates fewer carbon emissions than gasoline, its relative merits are widely debated. Some worry that sugarcane cultivation for ethanol may be responsible for deforestation in places such as Brazil. Others say Brazil's ethanol industry (which now supplies a third of the fuel being used by vehicles) has improved air quality and generated national growth. Some argue that corn-based ethanol takes more energy to produce than it provides and scientists have shown that, when burned as a fuel, ethanol generates less power than other hydrocarbon fuels such as octane and propane. Yet, ethanol's proponents say that the fuel is 'carbon neutral', since the carbon absorbed while the crops are growing offsets the emissions produced when it is powering vehicles. Some point out that ethanol competes with the food supply and could push up the price of agricultural products. Others believe that, given the right technologies and trade policies, the crops could provide a valuable source of revenue for farmers in developing countries.

Meanwhile, analysts cite cellulosic ethanol – which uses

plant wastes, such as grasses, corn stalks and wheat straw – as a more economically viable alternative to corn or cane, if efficient ways of breaking down the plants' cell walls can be found. Not only is cellulosic ethanol thought to produce several times more net energy than corn-based ethanol, it also generates a byproduct called lignin, a fuel that can be used to power bio-refineries. Moreover, cellulosic ethanol cannot be accused of competing with food cultivation, since its raw materials are waste matter and can be grown on land that is fallow or used for grazing. In Mongolia, then, as the ancient moveable lifestyle disappears, perhaps the country's vast grasslands might be used to grow the plants needed for cellulosic ethanol. If so, the grasses of the steppe would not only provide a source of income for the nomads. They would also generate a new kind of energy – a cleaner fuel with which to help deliver dinner.

8

Barrels and Bouquets

The oak tree leaves its mark

Wine: Alcoholic drink produced by fermenting the juice of grapes

Origin: Caucasia and Mesopotamia around 6000 BC

Etymology: From Old English *win*, derived from the Proto-Germanic *winam*, from Latin *vinum*

Legends: Wine is said to have been discovered when one of the consorts of Jamsheed, an ancient Persian king who kept his grapes in jars, mistakenly drank from a jar that contained spoiled fruit. She fell into a deep sleep and when she woke pleasantly refreshed, she reported her experience to the king. Jamsheed ordered that his next batch of grapes should be fermented

*I*N 1805, AN ILLUSTRATED children's book called *Little Jack of All Trades, With Suitable Representations* was published in London. It was designed to enlighten young readers about the various trades that were common at the time, from butchers and bakers to tailors and carpenters. Its popularity crossed the Atlantic to America, where versions appeared in the 1850s. The introduction certainly gives readers a rosy picture of the world. 'Society,' we are told, 'resembles a bee-hive, where, in producing a store of sweets all are employed – all live cheerfully – and whilst each individual works for the general good, the whole community works for him.' As well as such charming interpretations of the world of commerce, the book contains a series of poems, accompanied by short explanations of the importance of each trade. Among them is a section called 'The Cooper', and it starts with the poem:

In ages past, to hold wine fast
And keep it good and long,
They made a bag with skin of nag
And sew'd it neat and strong.
But now-a-days, the Cooper lays
The staves so close together,
That with his cask, it is no task
To keep out every weather.
Then, if you need work done with speed,
You'll find me very handy,
To make a tun that will not run,
To keep your ale or brandy.

Barrels, we are told, transport and store everything from beer, wine and cider to sugar and oysters – 'nothing answering the purpose so well as barrels, to hold certain goods, liquid and dry'. In this statement, the writer was spot on. The barrel is an ingenious and sophisticated piece of engineering. Constructed without nails or glue, it is a watertight container that is extremely tough. Moreover, its bulging shape means that, when on its side, less than a square inch of surface area touches the ground, making it easy to roll – a barrel holding up to one hundred gallons can be moved with one hand. The barrel is one of the world's greatest transport inventions, with a design so successful that barrels constructed today are little different from those made by the Bronze Age craftsmen thought to have invented the device.

However, the barrel has another remarkable talent not mentioned in *Little Jack of All Trades*. It can change the taste of the food and drinks stored inside it. For winemakers, oak has become a powerful aid to creativity, generating an astonishing range of flavours and textures. Everything from the climate in which the forest has grown and the grain of the tree to the way the wood is seasoned, aged and toasted contributes to the complex palette of flavours from which the winemaker chooses when designing a wine's bouquet. Analysing these bouquets has become high art that, to the uninitiated, sometimes seems to border on the ridiculous. 'Flowery peaches', 'black cherries' and 'redcurrants' are comparisons that seem sensible enough. Then there are the more unlikely 'pencil shavings', 'tar' and 'party balloons'. Perhaps the most bizarre descriptions to have entered the lexicon of the wine aficionado are 'petrol' and 'rubber'. The comparison with a transport fuel and something used in car tyres is strangely appropriate. Transport technology may have influenced politics, economics and culture but it has also had a profound effect on the taste of some of the things we eat and drink.

For centuries, mankind has experimented with different techniques to make food taste better. Creating new foods, as Mongolian nomads understood, involves chemistry, biology, a good deal of patience and often the right sort of container. In the constant search to add richer and more sophisticated flavours to foodstuffs, all sorts of boxes, pots and bags have been deployed. Raw products are often left for substantial periods of time to allow them to soak up characteristics of whatever they have been packed into.

Take thousand-year-old eggs. While these Chinese delicacies are not actually abandoned for a millennium, duck, chicken or quail's eggs are left for anything from a few weeks to a few months, coated with a mixture of clay, ash, salt and lime, packed into rice straw and buried. When they emerge, the eggs have been transformed. The white has turned into a brown, translucent, tasteless substance while the yolk takes on a rich green colour, a creamy texture and a powerful cheese-like flavour. It is perhaps not surprising that the eggs develop a taste resembling cheese. Cheeses can also be matured for periods of several years, wrapped in cloth or contained within anything from fig and vine leaves to clay pots and goatskin, and then left in caves or cellars.

However, the containers used to perform these culinary alchemies do not have the barrel's double purpose. For the genius of the barrel is that not only can it change the taste and texture of the foods and drinks contained within it. It also makes them highly moveable. It is the perfect marriage between high art and utilitarian function, cuisine and cargo, beauty and strength. And the barrel is certainly strong. Indeed, its sturdy form is based on architectural principles. In barrel-vaulted buildings, buttressing contains the outward thrust of the stones in each of the vault's semi-circular arches. When used on a barrel, the circle is completed. Hoops contain the outward energy of its components, the staves, and each stave serves the same function as the keystone in an arch (something that becomes clear when looking at a barrel from above). The radial rays of the oak, the wood used for most barrels, combined with these robust architectural principles, make the barrel an extraordinarily durable container.

Constructing a barrel is no easy feat. The cooper's job is to bind a series of wood staves together with metal hoops. Getting the staves into the right shape is the first task. Each needs to be tapered at its ends. It must have exactly the right curve and the edges must be cut at exactly the right angle so they fit snugly next to their neighbours. Open flames or steam are used to bend the staves to create the barrel's bulge. Then, using drawknives – two-handled knives with curved blades – the cooper hollows out the inside of the stave and gives it a convex shape on the outside, shaving off thin slivers of wood until it is just right.

Once all the staves have been made, they need to be assembled – another fiendishly difficult job. A windlass, a sort of winch, pulls the staves together so a temporary hoop can be placed around them. If the staves have been correctly fashioned, the tension will force them tightly together. The barrel, now known as a 'gun', is left to cool and dry. Grooves are cut into the ends of each stave and the 'heads' (circular discs at each end) are given bevelled edges and hammered in place. Permanent hoops are bound on to the barrel and what is now a 'wet' or 'tight' barrel ('dry' or 'slack' barrels carry non-liquid goods and do not require the same precision) should be watertight. Acquiring barrel-making skills takes years of practice, strong arms and an intimate knowledge of wood. No measuring tools are used. The cooper does everything by eye and experience. Today, the master coopers who supply wine-makers are highly paid craftsmen whose barrels sell for thousands of dollars – living craftsmen continuing to thrive from a process that dates back millennia.

Early barrel makers were not considering how to change the taste of food when they started to construct curved containers from planks of wood. They were looking for something in which to move liquids about without incurring leaks. Some speculate that wooden boats inspired the first barrels – instead of keeping water out, the same construction principles could be applied to the task of keeping liquids in. This may explain why the word 'vessel' means both a container and a ship. Appropriately enough, France was in the picture from the barrel's earliest moments. It is thought that Celts from Gaul devised the first barrels as early as 3000 BC. The Celts were skilled technicians and experts in the use of wood and iron. Historians have little to go on, but they believe these craftsmen constructed barrels with wooden staves bound together by iron hoops that closely resembled today's barrel. It is a plausible theory. After all, the Celts were known to have helped the Romans construct some of their more ambitious stone vaults.

In any event, by the end of the second century AD, the Romans were replacing their heavy earthenware amphorae with lighter wood barrels. Unlike the amphora, the barrel left no physical reminder of its presence in the Roman era.

However, it does make an appearance in Roman art. Among the carved relief images spiralling up Trajan's Column in Rome, which was dedicated in AD 113, is a scene in which boats are being loaded with casks that would have contained wine, olive oil, vinegar or fish sauce. Throughout the Middle Ages, the barrel remained a crucial transport mechanism, helping supply the crusaders during their expeditions to reconquer the Holy Land from the Muslims. In his twelfth-century chronicle of the crusade led by Louis IX, Jean de Joinville describes with admiration the 'great barrels of wine' amassed in Cyprus in advance of the king's arrival, which 'were stacked one upon the other in such sort that when you looked at them in front, the stacks seemed as if they were barns'. In late fourteenth-century Italy, barrels were essential agents in the burgeoning mercantile trade that funded the art of the Renaissance.

By the time the poem in *Little Jack of All Trades* appeared, large areas of the world were in the hands of a global powerhouse – the British Empire. As the Romans had demonstrated, with global empires comes the globalisation of food. This was certainly true for nineteenth-century Britain. In the explanation following the poem, *Little Jack of All Trades* told its readers that the cooper's trade 'enables us to send a variety of articles from all over the world'. The barrel enabled the growing international exchange of food to take place. Its job was not to conjure up bouquets or aromas – it was too busy moving around the world's feasts. Coopers were being engaged in the hundreds to produce barrels on an industrial scale. For, if the shipping container has become modern globalisation's pre-eminent transport tool, the barrel occupied a similar position in the nineteenth century.

Barrels brought sugar from the West Indies, and sent it out again to be sold in continental Europe. Rice came from South Carolina in barrels, which also carried cod from New England. In New England, molasses arriving from West India were distilled into rum and sent to Africa in barrels where the liquor was traded for slaves. Supported by global commerce, the British Empire was creating a commercially interconnected world that stretched from the Caribbean to Asia. India was the jewel in the empire's crown, but not only in political terms. 'People now at last begin to view those Indian affairs, not simply as beneficial appendages connected to the empire,' wrote Thomas Pownall, colonial governor of Massachusetts, in 1773, 'but from the participation of their revenues being brought into the very composition and frame of our finances; from the commerce of that country being indissolubly interwoven with our whole system of commerce.' This system of commerce underpinned the success of the British Empire – and was supported by the sturdy barrel.

Over at the West India Docks, hundreds of coopers toiled throughout the day, hammering furiously at hoops and staves of wood to keep up with demand for barrels. 'Here we had much work, having such great importations of wine and rum,' wrote William Hart, an English cooper who in 1803 started working for the West India Dock Company, an employer with whom he was to spend the next thirty years. Hart, born in 1776 and raised by his grandparents after his parents and siblings succumbed to disease, described the hardships of his trade in a memoir that offers historians a rare glimpse into the life of a cooper working in London at the turn of the nineteenth century.

As well as being physically demanding, the job was highly

competitive. When Hart's grandparents were seeking an apprenticeship for him, a cooper in the town of Leighton turned down their request, explained Hart, 'lest I should set up against him'. Eventually, a man named Wingrave agreed to take him on, and he was bound as an apprentice for six years. It was a difficult time for the boy. At just twelve he was, he wrote, 'very little for my age and weakly also, of a timid disposition and not one person in the place that I knew. I was away from all friends and acquaintances and very unfit for a business that required strength and energy.' It was a sentiment no doubt experienced by hundreds of his fellow apprentices as they prepared to start a lifelong relationship with staves and hoops.

In the late eighteenth century, artisans were converging on London from all over the country to find work in what was then a commercial boom town. Hart was among them. He had been impressed by the large cooperages he saw at Shadwell when visiting his uncle in 1794. 'Seeing so many men at work and getting great wages excited me to come to London,' he wrote. At the time, Britain's capital was the world's largest port, alive with activity as goods arrived from all over the world. New dock facilities were being built on an unprecedented scale. The new developments ripped the East End apart. As Hart recalled, 'whole streets were pulled down to make the London Docks'.

The new infrastructure could hardly have been more different from the old system, where vessels crowded into the Thames, jostling for position in a massive maritime traffic jam. Instead, well-ordered quays, smart transit sheds and immense warehouses became the logistical transit points for the empire's trade. Sitting on the Isle of Dogs peninsula, the West India Docks, completed in 1802, constituted one of the country's most impressive civil engineering projects. In 1805, the London Docks opened, followed by the East India Docks in 1806. These were the equivalent of the vast container terminals seen today at ports such as Singapore, Rotterdam, Hamburg and Los Angeles, where steel shipping containers are stacked up in their thousands.

Unlike containers – most of which are now manufactured in China – barrels were constructed on site. Each had to be made by hand and, when full of liquid, required constant surveillance and regular tightening to prevent leaks. Hart, who was then working among the wines and spirits, often had to make huge casks holding 180 gallons that were 'very stout for ships'. He was responsible for making and maintaining a huge number of barrels. Hart recalls a year in which there were 47,000 casks of wine and rum at the dock, 12,000 of which were under his charge. Like his fellow coopers, Hart was a humble craftsman. Yet his trade played an essential role in the development of food transportation, helping change society's tastes as new and exotic foodstuffs were shipped across the globe.

So how did barrels, once workhorse transport vessels, come to play a more creative role as tastemakers? It remains unclear. Olivier de Serres, an early-seventeenth-century agriculturalist, writes about fermentation of wine in oak tanks. Yet it was only in the twentieth century that the science behind this alchemy was established. Since then, vintners have been experimenting with the different ways they can harness oak in the wines they produce. They have plenty of variables to work with. For a start, there is the tree itself – French, American, eastern European, or Baltic forests are the most common sources. Each passes a distinct flavour on when made into barrels. French oak is generally preferred for the finest wines while American has a more assertive oak flavour. However, before the planks have been anywhere near the cooper's tools the wood is allowed to age at the stave mill, and different flavours accrue to it, depending on the climate in which it is left to rest. Another choice is whether to 'air-dry' or 'kiln-dry' the staves (each producing a different effect). When the cooper turns the aged wood into a barrel and it is held over open flames to 'toast' it, another set of variables is thrown into the mix in the toasting levels (medium toast, medium-plus toast and heavy toast are among them). Moreover, every cooperage has a different interpretation of these levels. Some even deploy sophisticated computer software to calculate the exact heat and time of the toasting.

In short, the barrel provides an extraordinarily varied spice rack of flavours from which to choose – and that is before the grapes have even been harvested. Once the oak has been turned into a barrel and is sitting quietly, full of wine, in the vintner's cellar, the wood really starts to work its magic. Tiny pores in the oak allow slow oxidation to alter each of the various constituents of the wine – sugars, tannins, pigments – forcing them to react with each other, creating liquid with a wonderfully soft, supple character. The maturation then continues in bottles, although with micro-organisms deprived of oxygen (no air can enter the bottle through the cork), this process is much slower.

Alcoholic drinks such as wine, rum, whisky and port – as well as products such as balsamic vinegar – have all had their flavour altered by this eminently moveable wooden vessel. Nor are the peregrinations of the oak tree the only example of food acquiring the redolence of the road. Take Russian Caravan Tea. Several explanations have emerged as to how the tea – named after the camel caravans that in the eighteenth and nineteenth centuries moved tea cargoes from China to Russia – acquired its distinct smoky flavour. One delightfully romantic theory is that the pungent aroma of the camels carrying the sacks of tea permeated the contents of their load. Another is that the tea's flavour was acquired through exposure to smoke from the campfires lit by the tea traders every night. The tea certainly had plenty of time to undergo its transformation – it took at least half a year to make the 6,000-mile journey from the Chinese border to Russia, and the voyage was harsh.

A less poetic theory appeared in *The Dublin Review* of 1888, where an account ascribes the taste of Caravan Tea to its exposure to climate. In a passage that discusses the debates among the merchants about the possibility of a railway across the Russian steppes, the author writes:

> The southern route by Odessa is far cheaper, but the tea is
> supposed to suffer in flavour in its transit through the tropical
> seas, while it improves in its passage through the cold dry
> climate of Mongolia and Siberia, by losing that unpleasant
> taste of firing [whereby tea was dried using direct heat]. As
> Russian epicures believe that a peculiar delicacy of flavour is
> imparted to it by the slight moisture it absorbs when nightly
> unloaded and placed on the snow-covered steppes, the
> enhanced price it commands compensates for the greater
> expense and difficulty of its carriage by this route.

In 1918, yet another theory emerged in the description by Raphael Pumpelly of a conversation he had with a tea trader called Phillippeus whom he met while travelling in Siberia in 1865. His fellow traveller evidently suggested another reason for the tea's flavour. Caravan tea, Phillippeus had explained, 'was from the same district of central China and from the same quality leaf as that sent by sea to Europe and America. The real difference was in the curing; the tea going by caravan overland in winter was "fired" only once, while for that going by ship through the tropics three firings were necessary, and these diminished the flavor.'

Whichever story is more accurate, it seems clear that climate was at least part of the answer to the mystery of Caravan Tea's heady aroma. But if Russian traders tried to avoid exposing their tea to heat, others found voyages through tropical temperatures actually had a beneficial effect on their goods. For one product, it all started on a small piece of land 400 miles west of Morocco. The island of Madeira was discovered by Portuguese sailors in the early fifteenth century and shortly after was claimed by the king of Portugal. Just 30 miles long and less than 14 miles wide, it soon became a profitable part of the country's empire.

Madeira has a pleasant, sub-tropical climate, with a natural beauty that attracts foreign tourists. It is overlooked by a range of craggy, mist-shrouded mountains whose lower slopes are covered with the terraced vineyards that for centuries have produced the grapes used in the wine to which the island gives its name. The fortified wine known as Madeira is thick and sweet, and is held to possess an aroma of caramel, crème brûlée or burnt molasses. The association with burning is close to the mark – Madeira has been, literally, 'cooked'. In a process that causes French vintners to wince, the wine is kept warm until it has matured. It is a highly irregular practice in the world of wine making – and its origins lie in transport.

Sugar fuelled the island's economy until the late sixteenth century when, after Brazilian and Caribbean cane started to dominate European markets, landowners turned their farms over to vines. The wines of Madeira soon became well known and by the seventeenth century they were being shipped not only to Portuguese possessions such as Brazil but also to British colonies such as America and the West Indies. To help them endure the long voyage across the Atlantic, wine merchants started adding a little brandy to the barrels.

They held up remarkably well, and the brandy, combined with the equatorial heat, appeared to enhance the liquid, producing a richly complex drink achieved elsewhere only after several years of maturation. Some even believed that the rocking motion of the vessel played a part in improving the wine's condition. So potent was the taste and texture of this new wine that during the nineteenth century, merchants would actually load their wines on to vessels and send them on long trips across the Equator and back before making them available for sale (Madeira's Portuguese nickname is *vinho da roda* or 'wine of the round voyage').

Today, it is impractical and far too expensive to indulge the wine in such journeys. So, the producers of Madeira have come up with a way of replicating the old tropical voyages of their wine. It is called *estufagem* (an *estufa* is a Portuguese hothouse). The fermented wine, fortified with brandy, is heated either by leaving the casks under the eaves of the roof where the sun bakes them or, in the more industrial version, by putting it into large tanks where it is heated by coils or pipes to about 45° Celsius. The liquid is kept at that temperature for about six months, after which it is allowed to cool slowly before being stored in oak casks. The best Madeira wines are highly prized

– yet it was a chance discovery made by mariners that led to this eccentric vintage.

Much of what we eat and drink has emerged as the result of such happy accidents. As in the case of Madeira, most of them took place long ago. Today, with new tastes and textures carefully fabricated by the food industry, it is rare that a genuinely new product emerges by chance. One exception is a drink that was born in Scotland in 2002. William Grant & Sons, the Scotch distilling firm, were working on the launch of a new whisky – one that had been finished in ale casks. The idea was that the beer-flavoured barrels would give the whisky a rich, malty taste. Grant's took a beer produced by Edinburgh-based Caledonian Brewery and left it in American white oak bourbon barrels before removing it and replacing it with whisky. Once it had worked its magic on the bourbon casks, the beer was thrown away.

One day, the distillery manager at Grant's, Mike Webber, rang his colleague at Caledonian, Dougal Sharp, and suggested he come over and try some of the beer being discarded by the whisky maker. It was delicious – infused with vanilla, toffee and orange aromas. Seeing the potential of the drink, Sharp approached Grant's, with a proposal to set up a separate company within the Grant's group in order to produce and market a new kind of beer – one matured in barrels bought from Kentucky bourbon companies. The distillery agreed to the scheme. Sharp, who had been planning to take a business degree, immediately changed his plans and started running the new company. 'It's a chance in a lifetime when you discover something new,' he says.

The idea of using barrels to age beer was a complete break with the past. Brewers, unlike their counterparts in the wine industry, used barrels strictly as a means of transport. In fact, beer barrels were steeped in water to remove any flavour before being used for transportation. New casks were lined with wax or enamel to prevent the beer coming into contact with the wood. 'My predecessors in the beer industry found oak was a real problem,' says Sharp. 'It was this step whereby the barrels had been used once for bourbon that made all the difference. The bourbon takes out all the flavours we don't want – the astringency and the really buttery oak character – and leaves the vanilla, slightly citrus, oaky character.' Now, after it has been brewed, the beer is left in the bourbon casks for at least thirty days. Then each barrel is nosed and tasted and, if deemed ready, blended in a 'marrying tun'. The final product is packaged in a chic whisky-style box. The barrels are then used a third time, for Grant's whisky. 'Being canny Scots, we don't like to see anything go to waste,' says Sharp.

Sharp believes his oak-aged beer, sold under the name Innis & Gunn, could be the start of an entirely new approach to brewing – one that, with its ageing, nosing and tasting, has a lot in common with wine making. The company has already tried putting different styles of beer in bourbon casks to add to its range of drinks. One of Innis & Gunn's newcomers, India Pale Ale matured in bourbon casks, also has its roots in transport. India Pale Ale was originally an export beer sent from Britain to India, where soldiers and colonial adminis- trators thirsted for a taste of home. However, getting the beer to them involved a six-month voyage through extreme tropical temperatures, which turned the drink flat and sour. In the early eighteenth century, brewers found that if they raised the alcohol level and added fresh hops to the barrels, the beer survived its long voyage to the Far East – and tasted pretty good when it arrived. India Pale Ale was born.

Because of the sheer volumes of wine drunk around the world these days, delightfully eccentric practices such as those of nineteenth-century Madeira wine merchants are no longer practicable. And in the same way as Madeira's producers found that the effects of moving their wine about in tropical oceans could be replicated by heating tubes, some winemakers no longer go to the trouble of crafting oak into a barrel at all. To generate the taste of wine matured in oak, companies have found a more efficient method: putting planks or large teabags full of oak chips into stainless steel vats of wine. Many French winemakers are, of course, appalled. However, as demand for wine at affordable prices has risen and heavily oaked varieties such as chardonnay have gained popularity, vintners have found alternatives to the expensive, labour-intensive practice of crafting barrels.

Among the pioneers of the new approach were a couple of Californian schoolteachers, Cole Cornelius and Bob Rogers, who spent their summer holidays in the 1970s making some extra cash by going round the vineyards of California to shave off the inner surfaces of their barrels (this reinvigorates the barrel after the wine has extracted all its flavour) and re-toasting them. Eventually they formed a company called Custom Cooperage.

Yet Cornelius and Rogers were not happy with the results of their labours. The recycled barrels simply did not result in wine with desirable bouquets. They decided to come up with something that could be put in an old barrel to give the wine the flavour of a new one. The barrel stave was born. The company, now called Innerstave, sells and installs a system of oak staves, arranged round the insides of barrels and kept in place by circular hoops, that recreate oak flavours in neutral barrels. As well as stave systems, oak chains can be threaded through the bunghole. Innerstave also sells all kinds of 'oak adjuncts' that have broken free of the barrel altogether: staves for tanks, bags of blocks, 'oakplus flour' (small particles of wood), wood chips – toasted or untoasted – and fines (smaller pieces of wood). For the vintner, this cuts costs. Not only are new barrels expensive – running into hundreds of dollars – but also their flavour only lasts about three years.

Purists still disapprove, as Alicia McBride, Innerstave's sales manager, will tell you. She says she has a hard time selling to traditional winemakers. 'I'll call up a winery and they will say: "We don't cheat – we only use real oak,"' she explains. 'So I say: "Oh, that's perfect – I only offer real oak." This is a marriage made in heaven.' Despite resistance in some quarters, a whole industry has emerged to produce these oak adjuncts, and it is no longer only the cheap chardonnays that benefit. Innerstave says it sells its products to winemakers who charge up to $150 for their bottles of wine. Even the Europeans are warming to the idea, particularly since the rules banning the use of oak adjuncts were relaxed in 2005. Many companies now offer a choice of wood from American, French or eastern European forests. In Napa Valley, Barrel Builders ('When the chips are down, come to Barrel Builders' is its sales line) offers to custom-toast its chips, which come in reusable nylon mesh bags. Moreover, these companies age and toast their oak products, giving them the same kind of obsessive attention that others would devote to the staves in their barrels.

One might be forgiven for thinking that this was the end of

the line as far as oak's ability to enhance the flavour of food and drink goes. However, at a stylish seafood restaurant in the heart of San Francisco's financial district, the tree's flavours have been taken still further away from the forest. At Aqua, Lionel Walter uses toasted French oak chips to flavour his crème brûlée. 'It's interesting – you get all the woodiness from the oak wood just transported into the cream,' says the French pastry chef, who lets the oak chips 'age' his hot cream for about ten minutes before baking it. 'You have the same tone you find in wine – a little bit of vanilla and then this toasted, buttery kind of flavour.' Oak chips designed to mimic barrel staves finding their way into a cream-based dessert? William Hart might be appalled at the idea – but perhaps he would be mollified if he were to taste a spoonful of Lionel Walter's 'oak-aged' crème brûlée.

Apart from rare exceptions such as Dougal Sharp's oak-aged beer, as well as the odd creation by an inventive pastry chef, the days when transport technologies inadvertently contributed to the palette of flavours in our foods are over. The great tankers that rattle across Europe carrying Spanish olive oil are meticulously cleaned and maintained, since olive oil is highly sensitive to rust particles and water. The US military now uses radio frequency identification technology to monitor the condition of the food in ration packs and measure any contamination or degradation. Food companies must comply with strict health and safety regulations when it comes to moving their products around. Take one look at the rules formulated by the European Union or the US department of agriculture's Food Safety Inspection Service and it becomes clear that what we eat is now permitted to travel only in the most sterile of containers, leaving little room for fortuitous enhancements to take place en route.

This does not mean that the food industry is giving up on devising new flavours for the things we eat – far from it. As with probiotic and fruit-flavoured yoghurts, most of our foods now have more to do with science than with culinary art. Products that are labelled 'banana flavour' or 'strawberry flavour' have probably had little or nothing to do with the offspring of the giant herb popularised by the United Fruit Company or the 'fruit of rosy hue' and 'modest grace' eulogised by Helen Maria Williams, the nineteenth-century poet.

Companies that design artificial flavours are part of a huge global industry. In the laboratories of these companies, food technologists have ingredients with names like isoamyl isovalerate, benzyl acetate and ethyl butyrate at their disposal and deploy 'predictive modelling tools' and 'high impact aroma building blocks' as they determine which of the tastes they have engineered in the lab will prove most popular out on the supermarket shelves. Of course, the alchemy being practised by these scientists has its roots in the kind of experiments still being conducted by winemakers in their search for new bouquets. These days, however, the development of new food flavours more often takes place in the laboratory – not in a simple, moveable wooden container.

9

A Quick Cuppa

Commercial competition speeds the racing tea clippers

Tea: Dried leaves of the tea bush (*Camellia sinensis*) steeped in boiling water

Origin: China and India

Etymology: From Malay *the*, and Chinese (Amoy) *t'e* and Mandarin *ch'a*

Legends: Tea drinking supposedly began when Emperor Shen Nung, a herbalist and scholar, was sitting under a tree waiting for some water to boil. Stirred by the breeze, a few leaves from the tree fell into the water he was about to drink. Tasting the liquid, Shen Nung immediately recognised its power to refresh and gave the brew his official approval

$\mathcal{I}$T IS 1873 AND, in a drawing room in Eaton Square, members of the Belgravia Women's Poetry Society have gathered for a reading. Packed into the dark interior is a confusing clutter of art objects, textiles, houseplants and furniture. Everything is highly ornate, from the Gothic commode on one side of the room to the large lamp with a cranberry-coloured glass shade made by Benetfink & Co. of Birmingham. Tables and chairs are draped with tasselled cloths, antimacassars and velvet runners. On a side table, ferns sit trapped within a series of miniature glasshouses and on the mantelpiece are some fine Staffordshire figurines. Many of the ornaments and pieces of furniture betray the influence of the Far East. Right now, the ladies are admiring a large Sèvres vase adorned with storks and exotic grasses. The vase sits on top of an ebonised mahogany table, clearly inspired by Japanese woodwork, and the wallpaper imitates an elaborate oriental textile.

While pleasantries as extravagant as the décor are batted about the room, the maid enters with a tray of hot water, a tea set and – the centre of attention – a walnut veneer tea caddy. Lady Templeton, the afternoon's hostess, takes a small key, opens the caddy and spoons some of the precious leaves it contains into a teapot, which the maid then fills with boiling water. Piping hot tea is poured into porcelain cups, which are handed round. 'It came from *Thermopylae*,' says Lady Templeton, with a self-satisfied smile. Her remark has the desired effect. An awed silence falls across the room. 'Oh, goodness!' pipes up one of the younger, less restrained guests. 'But how on earth did you get it? I heard supplies had run out long ago!'

This little scene is invented, of course, but in fashionable sections of London society, such exchanges would have been common in the nineteenth century. For *Thermopylae* was not a department store or an exclusive tea boutique. She was one of the era's great racing clippers – the fastest cargo sailing ships the world has ever known. In 1872, *Thermopylae* narrowly beat *Cutty Sark* on the route back from Shanghai to London. It was one of many clipper races that caught the attention of the British public. In a brief but glorious couple of decades between the 1850s and the 1870s, these magnificent vessels would set off from the east coast of China, their holds packed with chests of the new season's crop of tea, and race back to England – all hoping to be the first to arrive home.

These crack sailing ships were daring feats of marine engineering. They literally 'clipped' the top of the waves as they sped across the oceans, powered by the natural, free and endless supply of wind. The clippers were among the most technologically advanced machines of the era, designed to achieve previously unimaginable seaborne velocity. Their captains – figures such as John Keay and Donald McKinnon – were among the celebrities of the day and their crews were among the world's highest-paid seamen. Shipbuilders and engineers honed the designs of these vessels while sailors dedicated their lives to them (and many of those lives were lost) – all so that tea could be moved thousands of miles across the oceans at ever-faster speeds.

Tea. It is the most English of beverages. For the British, a cup of tea is the answer to everything, from a welcoming brew on the arrival of friends to a source of comfort in a crisis. In travel guidebooks to the UK, writers frequently throw the phrase 'a nation of tea drinkers' into their introductions to the country. The British have precise instructions for making this seemingly simple combination of dried leaves and hot water, including the insistence that the pot (preferably porcelain) must first be warmed to ensure the water remains at boiling point when it meets the leaves. They even debate whether milk should be added before or after pouring tea into the cup – a question to which George Orwell gave his attention in a 1946 essay where he listed eleven stages in the making of a cup of tea. 'The milk-first school can bring forward some fairly strong arguments, but I maintain that my own argument is unanswerable,' he wrote. 'This is that, by putting the tea in first and stirring as one pours, one can exactly regulate the amount of milk whereas one is liable to put in too much milk if one does it the other way round.' Rarely has a nation been so obsessed by one drink. Yet this quintessentially English beverage came from the most un-English of places: China.

In the nineteenth century, China was ruled by the Qing dynasty. The Qing had been extremely successful in the conquest of new territories and consolidation of the empire, but it was also to be the last regime to rule in a millennia-old system of government. While western merchants had been engaging in trade with the country since the sixteenth century, China's rulers remained closed to outside influences, uninterested in what the west had to offer and reluctant to embrace a world that, with the help of trade, transport and modern communications such as telegraphy, was rapidly industrialising and globalising. Pressure from foreign powers and internal rebellions would eventually precipitate the collapse of the Qing, paving the way for the creation of a republic in 1911 and, in 1949, Maoist communism.

So what mental picture did the members of the Women's Poetry Society possess of this far-off 'Celestial Empire', the source of the fragrant liquid they happily imbibed? While today's marketers conjure up romantic images of oriental-looking hills and tropical plantations to help sell their teas, Lady Templeton's was kept in a wooden caddy. Nevertheless, the trappings of Chinese culture surrounded the ladies of Eaton Square. For a start, the Sèvres vase that attracted so much admiration was decorated in the Chinese manner. Lady Templeton's collection of ornaments included a stereoscope showing images of 'The Picturesque Land of Confucius' and 'The Chukiang River, Canton, with Its Enormous Floating Population'. By placing two photographs, shot at different angles, side by side, the clever device displayed scenes from China with a miraculous illusion of depth. Moreover, the ladies were drinking tea from a Wedgwood willow pattern service whose wispy blue decoration tells the story of a mandarin's daughter who, forced into betrothal to an older noble, elopes with her father's secretary, who is then captured and killed. The legend, ending with benevolent gods transforming the couple into doves, was at the height of its popularity in the Victorian era.

Immense physical effort and complex political struggles lay behind the delivery of the fragrant brew that would fill the Victorians' willow pattern teacups. However, for the ladies of Eaton Square, only one aspect of the journey of their tea was of interest – the clipper race that had delivered it. It was an event not unlike the once-fashionable competition to secure the first bottles of Beaujolais Nouveau, except that each clipper race lasted for about three months, rather than a matter of hours. Cash prizes were awarded to captains of the victorious vessels and huge bets were placed on the outcome of the races. It was thought that the freshest and earliest-picked leaves created the best brew, so merchants in London could charge a premium on each pound of the newest teas of the season, especially if they arrived in a famous vessel such as *Thermopylae*. No wonder Lady Templeton kept her tea under lock and key.

The most celebrated of the China clipper races took place in late May 1866, when a cluster of ten vessels gathered by the docks in Foochow (roughly opposite Taiwan in China's south-eastern Fujian province), their holds packed with thousands of chests of the season's first flush of tea. To the sound of cannons, they set off. This was no ordinary collection of cargo carriers. The vessels that departed from Foochow that day not only represented the most advanced sailing designs of the era but also displayed the most refined of aesthetics. *Ariel* was a particularly fine example. Her captain, John Keay, describes her in the most affectionate of terms. He writes:

> Ariel was a perfect beauty to every nautical man who ever saw her: in symmetrical grace and proportion of hull, spars, sails, rigging and finish, she satisfied the eye and put all in love with her without exception. The curve of stem, figurehead and entrance, the easy sheer and graceful lines of the hull seemed grown and finished as life takes shape and beauty . . . I could trust her like a thing alive in all evolutions; in fact she could do anything short of speaking.

Keeping this dazzling creature company were vessels such as *Fiery Cross* (one of the most successful clippers of the day), *Taitsing*, *Serica* and *Taeping*. Astonishingly, three of them – *Serica*, *Ariel* and *Taeping* – arrived on the same tide into

London, *Ariel* and *Taeping* docking within minutes of each other. Ironically, the arrival of so much tea at the same time flooded the London market, driving the price down, but this fact is often forgotten in the retelling of a race that has inspired its share of purple prose.

Accounts of the early twentieth century paint quite a picture. In a 1930 volume, *The Old China Trade*, Foster Rhea Dulles characterised the 1866 race as one 'without parallel in the annals of sailing' in which 'continents served as the marking buoys'. In a blow-by-blow account not unlike those of today's television sports commentators, Dulles described how, at the Cape of Good Hope, it was *Fiery Cross* that was still 'in the van' with *Taitsing* trailing the entire fleet. 'Then coming up on the Azores, the *Ariel* jumped to the front, and the *Taitsing* passed the *Taeping*, the *Serica*, and even the *Fiery Cross*. Nearing the entrance to the British Channel the *Taeping* and the *Serica* crept up on the new leaders, passing both the *Taitsing* and the *Fiery Cross*, closing in on the *Ariel*.' Finally, in what must in nineteenth-century sailing terms have been a photo finish, *Ariel* arrived, followed closely behind by *Taeping* and, later that evening, *Serica*. 'While the *Ariel* was the first to cross the finish line,' Dulles continued, 'its eight minute lead was cancelled because the *Taeping* had sailed from Foochow twenty minutes later. Victory consequently went to the latter vessel. It had won by twelve minutes on a 16,000-mile course!'

Travelling at about 17 knots (equivalent to more than 20 miles per hour), the velocity achieved by these ships would have been thrilling to the ladies of Eaton Square. True, George Stephenson's *Rocket*, the steam locomotive built in 1829, hauled its train easily at 30 miles an hour. However, many people in Victorian England still crawled along in a horse and carriage. The outlandish new steam-powered motor cars were governed by the notorious Red Flag Act of 1865, which stipulated that a man had to walk in front of them with a red flag – hardly a licence for breakneck speed. Even the steamships entering service in the mid-nineteenth century found the clippers hard to beat. When the *Persia*, a Cunard steamship, won the Blue Riband prize for making the fastest Atlantic crossing without refuelling in 1856, its average speed was just under 14 knots, a pace that was easily being matched by the clippers. In an era when technology increased dramatically the potential speed at which humans could travel, the wind-driven clippers certainly held their ground.

Of course, by comparison with the speeds achieved by vessels in events such as the Whitbread Round the World Yacht Race or the America's Cup, the clippers' 17 knots might not sound like much. Multi-hulled super-catamarans now achieve more than 40 knots. These modern speed machines are, however, extremely lightweight fibreglass structures, not large ships of wood and steel weighed down with hundreds of tonnes of tea. For the astonishing thing about the nineteenth-century clippers is that while above water, all was a glamorous flourish of sails and seaborne swagger, below the surface of the ocean were deep holds where neat ranks of wooden chests sat stuffed with tea. Moreover, the genius of the clippers was not just their bursts of speed in a fair wind. Their real brilliance lay in the ability to continue sailing with only the tiniest breath of wind to power them, allowing them to 'ghost' gracefully along the ocean in the calmest of conditions. To do so, they relied on a captain with nerves of steel and a large crew of experienced mariners who had to battle the toughest of sailing conditions. The most important ingredient in the clippers' success, however, was luck. The finest sailing ship and the most able of captains would go nowhere without wind, and it was this element of chance that made the clipper races so thrilling.

The three vessels that swept into London at the dramatic end of the 1866 race all came from the Steele shipyard at Greenock on the Scottish River Clyde. A centre of pioneering nautical design, Robert Steele's shipyard was among those that, between the 1820s and the 1860s, were responsible for the most successful sailing vessels then plying the world's oceans. In fact, it was the Americans who had come up with the concept of the clipper. In the USA, the phrase 'Yankee Clipper' might have won new life in the 1930s as the nickname for a famous baseball player, Joe DiMaggio of the New York Yankees, but in the nineteenth century it was a type of ship – one that caught the eye of the British. Abandoning the long-held notion that cargo vessels should be designed with all eyes on capacity rather than speed, the Yankee clippers were tall and sleek with large sails.

The British soon set about refining the model. Shipbuilders such as Steele created vessels with finer lines, hulls whose knife-like edge sliced through the water and robust sterns that could counter the effects of a large swell on the roughest ocean. Then there were the sails – enormous billowing clouds of white fabric. *Cutty Sark*, one of the most famous clippers and the only one still in existence, carried about 32,000 square feet of sail. That would have been enough for artists Christo and Jeanne-Claude to wrap the exterior of the Chicago Museum of Contemporary Art (which they did in 1969) three times over. Shipbuilders such as Steele – in fierce competition with other shipyards – were constantly honing the design of their vessels to shave even more time off the route home from China.

Those who followed the clipper races were unaware that, even before it left China and embarked on its three-month race across the ocean, British-bound tea had been on an arduous journey. High up in the Bohea Mountains – a range of grotesquely formed limestone crags now known as the Wuyi Mountains – was where the mild, humid sub-tropical climate and fertile soil produced a crop that filled more than 500,000 tea chests a year. At a place called Zongghihian, the tea was blended and packed in large warehouses, or *hongs*. It was then packed into chests and suspended on bamboo rods on the shoulders of thousands of coolies, who set out on a back-breaking six-day journey along a rugged mountainous path. They ended up at Hokow, a trading centre where the tea was loaded on to boats and sent along canals to ports on China's eastern and southern coasts, where it would be stashed inside clippers bound for England. Tea shipped from Canton, for example, had travelled nearly 1,200 miles before it had its first whiff of the high seas.

Once on the ocean, it would travel another 16,000 miles before reaching British shores. From the ports at Foochow, Shanghai or Canton, vessels generally headed down through the South China Sea towards the Sunda Strait and into the Indian Ocean. From there they made for the Cape of Good Hope, on Africa's southern tip, and headed home up across the Atlantic Ocean, through the Doldrums, past the Cape Verde islands and either up the English Channel and through the Thames to London or past the coast of Wales to Liverpool. The trips home from China were dangerous, right from the start. Many vessels met their fate before they had even set out to sea. From the Pagoda Anchorage at Foochow, ships were forced to navigate the narrow Min River, where strong currents could push them on to the banks. There, stranded, they could be boarded by pirates and other gangs who would kill the crew and strip the ship of its cargo. Piracy was rampant out on the ocean, too. And if they managed to escape the clutches of these waterborne criminals, sailors faced severe storms and monsoons.

If they survived such trials, however, a great welcome awaited the clipper crews at home. In London, messenger boys raced down to Mincing Lane to announce the imminent arrival of a vessel to fat-bellied tea traders. Crowds gathered by the docks to see the elegant creature arrive. In Liverpool, where the ships could make their final passage up the Mersey under full sail and in clear view of the general public, great noise and commotion would greet the sight of these masters of speed. The races continued well after the cheering of the dockside crowds had died down. While tea brokers in Mincing Lane haggled over prices for a particular consignment, that consignment was swiftly removed from the belly of the ship. On 20 September 1864, for example, *Fiery Cross* arrived alongside St Katherine's Dock at 4 a.m. and by 10 a.m. the following morning 14,000 chests of tea had been unloaded. In an era well before the advent of gantry cranes, forklift trucks and the modern shipping container, this was an astonishing feat.

The clipper's moment of glory would not last long. The opening of the Suez Canal in 1869, connecting the Far East with the Mediterranean via the Red Sea, heralded their decline. Because of its expensive tolls, long passages of calm and unpredictable winds, the clippers were unable to navigate the canal, giving steamship transport the competitive edge and bringing to a close the races that had captured the imagination of so many.

The clipper races were about a lot more than sporting prowess. Profit, of course, was the most powerful motivation behind the evolution of such impressive vessels, and two political developments served to intensify the commercial competition. First, in 1834, the East India Company lost its monopoly on the lucrative China trade. Then, in 1849, the British government revoked laws ruling that only British ships could trade in its ports, allowing the Americans to join in. The race was on, and it was a global one.

Tea was a profitable business. In 1830, 30 million pounds of tea were shipped from China to Britain. By 1879, that figure had risen to 136 million pounds. From cheap black Congou to good green Gunpowder or superior young black Pekoe, Chinese tea was the most fashionable of drinks and, in the nation's crowded new tea rooms, tea drinking became a popular social event. The brew even acquired seductive possibilities. In a scene from Mary Elizabeth Braddon's sensational 1862 novel *Lady Audley's Secret*, Lucy Audley, the heroine, offers her husband's dashing nephew, Robert Audley, a cup of tea. Braddon wrote:

> Surely a pretty woman never looks prettier than when
> making tea. The most feminine and most domestic of all
> occupations imparts a magic harmony to her every
> movement, a witchery to her every glance. The floating
> mists from the boiling liquid in which she infuses the
> soothing herbs, whose secrets are known to her alone,
> envelop her in a cloud of scented vapour, through which
> she seems a social fairy, weaving potent spells with
> Gunpowder and Bohea.

Whether they were after 'floating mists' and 'potent spells', or simply craving a dose of caffeine, nineteenth-century Britons firmly established the country's reputation as a nation obsessed with tea. Historians have suggested that even Britain's temperance movement, which began in the 1830s, had an impact on consumption since, at high-minded meetings across the country, lashings of tea were invariably served up along with stern warnings against the perils of excessive alcohol consumption.

However, if the British were feeling smug about drinking tea to battle vice at home, on the other side of the world their sacred drink was tied up with an addiction to a far more dangerous drug: opium. This narcotic, a dried juice derived from the unripe seeds of the opium poppy, had long been known in China. It was first brought to the country by Arab traders and used as a medicine. However, when in the mid-seventeenth century tobacco smoking, introduced by western traders, became so prevalent that the Ming emperor banned it, the Chinese turned to another addictive substance. Using the tobacco pipe for their new drug, the users would place a small lump of opium in a receptacle, heat it with a candle and, while lying on a bed, smoke the vaporised morphine. Opium soon became the narcotic of choice – and it was highly addictive.

While some opium had been grown in China, much of it came from India, where the Mughals controlled the industry. Despite harsh penalties (including strangulation) for the keepers of opium shops and dens, and an imperial ban on opium smoking imposed in 1729, the Chinese authorities could not prevent foreign consignments of the drug arriving at the country's ports. When, in 1773, the British gained control of Bengal and Bihar from the Mughals, this valuable commodity was put into the hands of the British East India Company, which increased production. Even a ban by the Chinese in 1796 on imports of opium and exports of silver failed to stem the flow.

While sallow-skinned addicts languished in Chinese opium dens – some smoking as many as 200 pipes a day – tea turned up on British shores in ever-increasing amounts. So vast were the consignments leaving his country that Commissioner Lin Tze-hsu formed a picture of a nation utterly lost without its daily cuppa. In a celebrated protest at the British opium trade sent to young Queen Victoria in 1839, Lin wrote that, when it came to tea and rhubarb, 'foreign countries cannot get along for a single day without them'. The commissioner's rhubarb theory (which was that without regular supplies, Britons would die of constipation) might have been a little far fetched, but he was closer to the mark with tea. Mrs Beeton's famous 1861 guide to household management declared that 'the beverage called tea has now become almost a necessary of life'.

The trouble was that the British were running out of the silver needed to pay for this 'necessary of life'. Silver had been China's preferred currency since western merchants had first shown up on its shores. But when America had declared its independence in 1776 (an event in which tea, of course, played a role), supplies of Mexican silver were taken from British hands. Inflation, too, drove up the cost of the metal. The British needed something else with which to finance their tea, and opium provided the answer.

Never slow to capitalise on a commercial opportunity, the East India Company ramped up its opium operations in Bengal, where, at its peak, the industry employed more than a million workers. To avoid becoming directly embroiled in the sordid business of drug dealing, the company sold the opium it produced to independent traders at auctions in Calcutta. These traders took it on to China (also using clipper ships), where the drug was sold to Chinese agents in Macau or Whampoa who would smuggle it into the country and distribute it to the waiting addicts. For Hong Kong trading houses, such as the one founded by William Jardine and James Matheson in 1832, opium was a highly lucrative business. For the empire too, it was a crucial source of revenue. In 1832, the year's crop represented one-sixth of the gross national product of British India.

With plentiful supplies of opium at their disposal, the British now had a commodity that, as China's appetite for the drug grew, could provide the funds with which to purchase their tea. One addiction was being used to pay for another. 'The average Victorian simply chose to close his eyes to the opium traffic and pretend it did not exist,' writes historian John Evans. 'This same Victorian would look shocked and shrug off as "humbug" if he was told that opium financed the British Empire, provided his high standard of living, and paid for all the tea he and his countrymen drank.'

Of course, the drug was not unknown at home – far from it. Opium wrecked the lives of many Britons, including literary figures such as William Hazlitt and Samuel Taylor Coleridge, while the opium dens of London and New York counted growing numbers of visitors. However, the scale of consumption there was tiny compared to that of China, where addiction reached epic proportions. In 1838, according to one record, 40,000 chests made their way into the country. Such large volumes of the drug took their toll not only on the populace but also on the country's balance of payments as more and more silver left Chinese shores. The emperor made several attempts to ban all imports of the drug. In 1839, Commissioner Lin confiscated thousands of chests of opium and flushed the leaves out to sea. Not long after, the emperor slapped an embargo on the export of tea and rhubarb.

The British were not too worried about supplies of rhubarb. However, they were not prepared to give up on their opium-funded tea trade. War broke out between the two nations. In the ensuing conflicts, known as the Opium Wars and fought in the mid-1800s, Britain beat the Chinese into submission, forcing them to open ports such as Shanghai to foreign trade, to legalise the opium trade and to hand over Hong Kong, an island that was to remain in British hands until 1997. Once again, moveable foodstuffs had precipitated far-reaching political developments. And for all the daring and heroism of the clipper races, there was a dark underbelly to the British love of tea: the deaths of tens of thousands of Chinese opium addicts.

The diminishing importance of the opium trade to the empire's finances coincided, unsurprisingly, with a shift in the geographical source of tea that would bring about the end of China's pre-eminence as a supplier. Politics was behind the relocation of tea production. It dawned on the British that they did not have to rely on a country like China, one that inconveniently refused to be colonised and that, even more annoyingly, fought back when opium was foisted upon it. While the British had prevailed in the Opium Wars, the conflicts had exposed the dangers of relying on China as the sole source of tea. Far better, they reckoned, to grow it in a place where they were in charge of things: their empire. With about one-fifth of the world's land under British rule, there were certainly plenty of options when it came to selecting new places to cultivate tea.

In 1834, a Tea Committee had been established to examine the potential for growing bushes in India from Chinese seeds. The argument for growing tea outside China was becoming compelling. The Chinese had been resisting imports of opium and the East India Company had just lost its global monopoly. The committee found that the tea plant was indigenous to Upper Assam. After much debate, it was decided that Chinese tea would fare better than local varieties. George James Gordon, a committee member, was sent to China to collect seeds and plants and hire experts in the cultivation of the bushes. Then, in the 1850s – in between the two Opium Wars – the East India Company sent a British botanist called Robert Fortune to China. He left with 20,000 tea plants in his baggage and headed straight for India. Tea plantations were soon being established in places such as Darjeeling and Assam, as well as in Ceylon and British colonies in Africa. While the racing tea clippers were winning the admiration of people in London, China was losing its grip on the global tea market. It has never since enjoyed such a dominant role as a source of tea. The country now competes with countries such as Kenya, India and Sri Lanka, and much of what is grown in China never leaves the country.

Keith Chen likes to warm the imagination of his visitors with a cup of tea. It is not served in the most inspiring of settings. His office is in an unprepossessing trading complex on He Chuan Road, a traffic-packed artery running through an industrial suburb of Shanghai. The place is sparsely furnished. Desks and office equipment share space with a large sofa set and a low-slung coffee table. A large map of China dominates one wall; on another is a photograph of Prince Charles on a visit to Shanghai. A vase of plastic flowers is the only other decoration in this thoroughly businesslike interior. The atmosphere may be unromantic, but as Chen serves up cups of tea, magnificent aromas waft around the room – delicate fragrances that seem to belong somewhere far away from this office. In fact, they do. The teas that Chen brews for his guests are thoroughly well travelled. They were blended, packed and sent from England to China by Bettys & Taylors of Harrogate, a traditional family business established in 1886 in North Yorkshire and famous for its old-fashioned tea rooms and blended teas. In what Taylors likes to call its 'coals to Newcastle' deal, the company has been shipping its decorative caddies and teabag boxes to Chen, who sells them to fashionable restaurants and hotels in China.

Chen is passionate about what he does. 'Tea is for sharing,' he beams, gesturing wildly. 'After all, a teapot is for two.' With a round open face, the rotund Taiwanese-born businessman explains that his love for the food and beverage industry, in which he has spent his career, makes it tricky to lose weight. His eyes sparkle and he frequently runs his hand through unruly tufts of straight black hair as he talks – particularly when the subject turns to business. His excitement is understandable. His ambitions for Pacific Trading Management Corporation, the company he and his wife Vivian own, go way beyond that of being a middleman for tea buyers. Chen wants to develop a chain of cafés serving tea as well as coffee, soft drinks and simple meals such as pizza and pasta. The company already has a branch office in Beijing and, working with a local Shanghai bakery, he is even developing a range of tea-inspired desserts such as Earl Grey cookies and Yorkshire Tea cheesecake.

The journey of Chen's teas, which have travelled to China in a large steel shipping container aboard a giant cargo vessel, cannot begin to compare with the romantic and perilous voyages of the chests of Bohea and Gunpowder that were conveyed in the holds of nineteenth-century racing clippers. But their story is an intriguing one. For Chen's brews are at the leading edge of a curious reversal in the direction of the tea trade. While, in the nineteenth century, tea-laden ships were bound for Great Britain, today a growing number of shipping containers are showing up at Chinese ports full of 'English' tea.

Here again, the world of food is becoming smaller as transport allows consumers all over the world to satisfy their desires for the new, the exotic and the unexpected. In China, a nation whose economy is expanding at an astonishing rate, city sophisticates in places such as Beijing, Shanghai and Guangzhou are acquiring increasingly global tastes – and English tea has landed firmly on the list of chic desirables. Meanwhile, along the Bund in Shanghai (now a city with more skyscrapers than New York), former trading houses and banks that would once have looked out at the tea clippers as they set off for England have been converted into upmarket department stores housing Armani, Cartier and Aquascutum boutiques. Offices where shipping clerks once buried their heads in paperwork are designer restaurants.

The shipping clerks now work at a vast container terminal in the newly developed Pudong district of Shanghai. In a large control room at the port a prominent sign reads 'Sincere Service, Meticulous Control'. It sounds like something Mao Tse-Tung might have written but this port, now one of the world's busiest, shows how far China has come since the Cultural Revolution. Here in the control room, banks of computers manage the movement of the cargo arriving out on the docks beyond the glass windows. That is where shipping containers carrying the Taylors teas Chen has ordered will land, lowered gently on to the quayside by a giant crane. Globalisation is much in evidence among the gantries and vessels of the Shanghai docks – but then for a city that witnessed the birth of the clippers' trade, this is nothing new.

Neither, of course, is tea drinking in China. Like Britain, this is a country of tea connoisseurs who recognise a decent brew when they come across one. At the Yu Yuan teahouse you can capture something of the atmosphere of this ancient and sophisticated culture. Rising up on stilts from a small lake in the centre of Shanghai, the ornate teahouse looks rather like the pagoda on Lady Templeton's Wedgwood willow pattern tea service. With dancing roofs and red carved wooden balustrades, it is reached via a zigzagging bridge designed to keep away evil spirits (who are only able to travel in a straight line). Today the quaint-looking structure is surrounded by a sprawling shopping mall of air-conditioned boutiques, mobile phone shops and fast-food restaurants. Amazingly, given the throngs of tourists and shoppers all around it, the teahouse itself remains remarkably tranquil. Inside the creaking wooden building, at small tables in tiny alcoves framed by shuttered windows, fragrant teas such as chrysanthemum and jasmine are served in miniature cups by old men in traditional Chinese dress. It is a scene that looks much as it would have done hundreds of years ago.

Many of the teas served in the Yu Yuan teahouse are grown in Fujian province. There, the hills of Anxi country are covered with bushes that produce the oolong tea for which the region is famous. These teas travel relatively modest distances between the plantation and the pot. Not so for the Taylors teas being sold in China. Taylors imports leaves from all over the world, then blends and packages them in Yorkshire. Taylors supplies the Chens with a variety of these blends, such as its celebrated Yorkshire Tea, as well as English Breakfast and Earl Grey and more exotic products such as Moroccan Mint Tisane and South African Kwazulu tea. However, the best travelled of them all is the China Rose Petal tea. This tea is actually grown in China, where it is blended with rose petals and sent to the company's plant in Harrogate. There it is checked for quality, packaged up and sent back to China. By the time it reaches Chen's warehouse, Taylors' rose petal tea has – like the frozen salmon sent to China for filleting before heading back to Europe or America – been on a monumental round trip: to North Yorkshire and back.

Some of the new enthusiasts for English-style tea can be seen at tables of cafés and bars in Shanghai's old French Concession, one of the international settlements created after the first Opium War. Here, the unstoppable force of modern development has not yet eclipsed the relaxed continental European atmosphere of this leafy quarter. Art deco apartments and grand villas built by colonial traders sit behind ranks of trees on wide boulevards while fashion boutiques and chic restaurants add to the appeal of a neighbourhood that is rapidly becoming one of China's hottest property markets.

At the Fragrant Camphor Garden, a trendy establishment on Hengshan Lu, young couples decked out in designer desirables sit at stylish black tables while Buddha Bar music plays in the background. Waiters deliver dishes with names like Matsuzaka Beef Shaba Shaba and copies of fashion magazines and international newspapers are available on request. Outside on the terrace, teenage girls gossip beneath the shade of white canvas umbrellas and creative types wearing black shirts and steel-rimmed glasses discuss business plans through a stream of espressos and cigarettes. As well as coffees and juices, the menu features a long list of teas: lemon, fresh fruit, strawberry, kumquat and jasmine, as well as the more traditional Earl Grey, Taiwanese oolong and Korean ginseng. Fragrant Camphor Garden's extensive selection includes a number of milk teas – mint milk tea, pearl barley milk tea, rum milk tea, cinnamon milk tea, peanut milk tea and mung bean milk tea.

The *nai cha*, or milk tea, phenomenon is an odd one. While English tea is generally taken with milk, the idea of putting milk in tea is for the Chinese an intensely disagreeable one. For westerners, it would be rather like putting a lump of butter into a cup of coffee (which the Vietnamese enjoy) or slipping a piece of pork fat into a cupcake (something the Chinese, who do this to their moon cakes, actually find delicious). Chinese people often find it hard to digest dairy products, which have not traditionally been part of their diet. For the older genera-tion, the idea of tainting the purity of tea with a substance such as milk is sacrilege. This is changing. China's citizens are starting to eat foods such as yoghurt and cheese and some of them are putting milk in their tea. The Chens have latched on to the new trend, and in doing so Starbucks, which started opening coffee shops in China in the late 1990s, has unwittingly given them a helping hand. 'If you serve milk with tea to Chinese people, they don't like it, but they will accept "tea latté" because of Starbucks,' Chen explains. 'So we serve tea latté in a mug with different sizes like "jumbo" and "tall". We are taking Taylors tea and applying a trendy café concept to it. You'd be surprised – it's so popular.'

'Tea latté' in a large mug is not the only form in which fashion-conscious Chinese consumers like to take their English-style tea. Not far from the Fragrant Camphor Garden café, boutiques such as Simply Life and Sha Ping sell dainty porcelain teapots, complete with milk jugs, cups and saucers. Decorated with English roses, camellias or art deco patterns, they look like something that might have been produced by Royal Doulton or Wedgwood – that is, until you examine the base and find 'Hai Ge Le' and 'Dong Yang House' inscribed there. Commissioner Lin must be turning in his grave.

The tea being drunk on the 10.05 a.m. First Great Western express service to Truro is not being served from dainty porcelain pots – it is slopping about in polystyrene cups with plastic lids. The drinks were grabbed from a food outlet before the train pulled out of Paddington Station. First Great Western has packed people on as if it were a budget airline. It is a hot day and flustered passengers struggle to find their seats and stash their luggage. After a flurry of mobile phone conversations in which the vital message 'I'm on the train,' has been conveyed, things calm down a little. People tuck into their sandwiches and open up their newspapers. As the train plunges south-west towards Cornwall, the verdant landscape seems to have a calming effect on the passengers. London's urban sprawl quickly fades from consciousness and those lucky enough to have secured a window seat begin to gaze out at the world beyond reinforced glass and air-conditioning.

Outside is the lush, green countryside. It is June and

hedgerows are rampant with summer growth. White elder-flower blossoms fall across the thick greenery like a fresh sprinkling of snow. The foxgloves are out in force – a proud pink-clad army on the march – and wild rhododendrons give an exotic twist to the normally restrained character of the rural foliage. The train speeds past ancient church spires, Victorian Gothic station houses and fields dotted with freshly shorn sheep. It is the very picture of Englishness. Yet, as somnolence works its way through the carriage, the dozing passengers may be unaware that they are heading straight for a most unusual English phenomenon – the first commercial tea plantation in the British Isles.

Appropriately enough, the fledgling tea garden ('garden' being the term often used in the industry to describe a plantation) is owned by a descendant of Earl Grey. The Honourable Evelyn Boscawen is Earl Grey's great-great-great-great-great-great-great-grandson and he lives on an estate that the Boscawen family first occupied in the early fourteenth century. With its 4-mile-long driveway, his estate, Tregothnan, is certainly impressive. Translated from Cornish, 'Tregothnan' means 'The House at the Head of the Valley', and, sure enough, the seventeenth-century manor house – enlarged and improved in the 1820s – commands a splendid view towards the River Fal and the surrounding valley.

Around the house is a spectacular 40-hectare garden and arboretum where botanical rarities such as a 200-year-old cork tree and some of the world's largest magnolias (they soar to 60 feet) line grand walkways shaded by ancient trees. The practice of botanical collecting goes back a long way here. The family is the proud owner of what is probably the world's only surviving Wardian case – a kind of mini-greenhouse designed in 1836 by

Nathaniel Bagshaw Ward to protect the exotic plants then being brought back from around the world. Wardian cases played an important role in creating the wealth of non-native flora and fauna in Britain today (perhaps Robert Fortune used one when he took his Chinese tea plants to India). Tregothnan is also home to a collection of more than a thousand camellias which, when they were introduced to the estate in the nineteenth century, were the first to be grown in Britain.

It was the camellias that inspired the move into 'English' tea, since tea is made from the leaves of the *Camellia sinensis* plant. 'Two hundred years ago, we pioneered the growing of camellias from China – so this set us thinking that surely tea wasn't impossible,' says Jonathan Jones, Tregothnan's garden director and the man who came up with the idea of establishing a tea plantation on the estate. 'People think tea has to be grown on mountains,' he explains. 'The reality is that it has to be cool to get high quality. And the climate in Darjeeling is really similar to this part of England – cool, moist and damp – so we have things that are native to Darjeeling growing in profusion here.'

Jones is a fresh-faced Cornishman with a soft-edged Cornish accent. Like Keith Chen, he is a cheerful optimist, and for someone with the word 'gardener' in his job description, he has acquired a remarkable grasp of the business world. A computer-generated image on one wall of his office reveals ambitious aspirations for an 'International Tea Centre' that, he says, will be 'a kind of front-of-house for the Tregothnan brand'. He likes to quote Adam Smith, the pioneering eighteenth-century Scottish economist, and throws terms such as 'customer demand' and 'bottom line' into the conversation – that is, until he is teased and the serious corporate image gives way to infectious laughter. Jones is, however, deadly serious about making the new venture successful as a way of securing Tregothnan's long-term future. 'I got interested in knowing what was this mysterious world of business and how could we make it work for the estate,' he says. 'Until recently, we were not able to say the word "brand" around here. Now we're building one.'

So far, Jones's and Boscawen's business remains small scale. The venture suffered early setbacks, such as a severe gale that uprooted all the first tea bushes in the kitchen garden and deposited them on the other side of the garden's high red brick wall (astonishingly, they all survived replanting). Tregothnan tea is now being sold in stylish contemporary packets at Fortnum & Mason, the posh London department store, as well as in a shop in Seattle. Boscawen – who like many of his ancestors is a passionate gardener – has recognised the need to devote some of his energies to business if his family's estate is to secure an economically viable future. Initially, he was sceptical about the idea of cultivating tea. 'I said, "Are you bonkers!"' exclaims the quiet but forthright Boscawen, with a twinkle in his eye. However, Jones persuaded him it would work and, with a Harvard Business School course for owner-managers under his belt, Boscawen is now as excited about the project as his head gardener. 'I'm the biggest critic of products,' he explains. 'But I have to say, it's fantastic.'

Jones has done much of the legwork when it comes to the tea garden. In 2002, armed with a scholarship from the Nuffield Foundation, a charity that funds research and practical experiments, he set off on a voyage round the world. The trip took him 'pretty much wherever tea grows'. He visited India, Kenya, China and Sri Lanka, as well as newer tea-growing areas such as the US states of South Carolina and Washington, a plantation in northern Australia owned by the Japanese and one on New Zealand's South Island. All this travelling brought home to Jones the fact that it was Britain's obsession with tea that led to large areas of the world's cultivatable land being covered with bushes of *Camellia sinensis*. 'The funny thing about tea,' says Jones, 'is that it's British through and through – we've been connected with the jolly thing right from the start.'

If Jones and his aristocratic boss have their way, English-grown tea may soon cease to be such a curiosity. Deliciously scented and elegantly packaged, Tregothnan's blends will certainly appeal to those who want to increase the amount of locally grown produce in their kitchens. However, for tea, the most well-travelled of commodities, to be produced on English soil is an odd turn of events. For consumers, too, the idea of English-grown tea may seem as incongruous as truffles from China or 'Florida' oranges from Punjab. Yet perhaps it is fitting that this, the most English of products, is at last being cultivated in the most English of places: an ancient estate in Cornwall owned by a descendant of Earl Grey. After centuries of being shipped around the world in everything from racing clippers to modern cargo containers, here in Britain's far south-west corner tea has finally found its way home.

10

Food with Altitude

Jet planes dispatch a strawberry for all seasons

Strawberry: Fruit of the *Rosaceae* or rose family

Origin: In Greek and Roman times, the strawberry was a wild plant; the garden strawberry was first raised in eighteenth-century France

Etymology: From Old English *streawberige*

Legends: According to ancient German lore, strawberries symbolised children who had died young. On St John's day – when the Virgin Mary is said to accompany children who go strawberry picking – mothers who had lost an infant would take care not to eat any strawberries lest their children in Paradise should be deprived of the fruits

OUR TIMES A week, an unusual flight takes off from an airstrip in Houston, Texas. On leaving the ground, the plane soars up over the Gulf of Mexico to more than 30,000 feet before plunging back down through the clouds at more than 500 miles an hour. Powered by two massive turbofan engines weighing 14,500 pounds each, the plane makes up to sixty of these giant parabolic arcs. It is one of the world's most spectacular roller-coaster rides. The organisation operating this remarkable flight is NASA, the US space agency. The aircraft is a C-9, the military version of a McDonnell Douglas DC-9. With aircraft and passengers falling at the same velocity for the twenty or thirty seconds before the plane swoops back up again, the effect of weightlessness is created, allowing scientists to conduct experiments under conditions similar to those in orbit.

Unsurprisingly, on such a heart-stopping voyage it can be hard to keep food down. Ten to twenty per cent of passengers on the plane throw up (NASA calls this the 'kill rate') before the two-and-a-half-hour flight is over. This might not sound too bad – until you hear that, during the forty-one-year career of the original KC-135 turbojet (the aircraft that was replaced by the C-9 in 2005), its crew had to mop up nearly 300 gallons of vomit, giving the plane its much loved nickname – the 'Vomit Comet'.

Regurgitated food is not the only side effect of space research. Sending supplies far into the atmosphere to feed astronauts has produced advances in food technology. Working on astronauts' diets, NASA developed irradiated meats that last several years without being refrigerated, for example. More recently, however, the prospect of travelling much further, to Mars, has given scientists at the Kennedy Space Center in Florida a new challenge. Among the more unusual ideas they are pursuing is the possibility of growing fresh produce on the journey. Referred to by NASA as the Bioregenerative Life Support Systems project, the idea is that fresh fruits or vegetables grown during the voyage could enrich the astronauts' diet. Scientists also reckon that doing a bit of gardening in space might be therapeutic, particularly when the astronauts' own backyards are millions of miles away.

Among the produce that may boldly go where no other produce has gone before is the strawberry, chosen because it can be induced to flower under the low light conditions present in spacecraft. Strawberry flowers grow outside the main structure of the plant, making it relatively easy to pollinate them by hand or mechanically in the absence of bees. What is more, antioxidants found in strawberries can, says NASA, 'reduce the cellular oxidative stress [cell damage] astronauts may face from cosmic radiation once they leave Earth's protective magnetic field'. Scientists are even considering the possibility of growing strawberries on Mars itself, with talk of cultivating what have already been dubbed 'Marsberries' on the Red Planet.

Helen Maria Williams, a nineteenth-century British poet, would have been appalled at the thought of strawberries growing inside a space rocket. Williams was a remarkable woman, translating literature, chronicling the French Revolution and welcoming many of Europe's most influential thinkers to her Paris salon. She also wrote a sonnet called 'To the Strawberry'. Her short poem captures the ephemeral nature of the succulent berry prized by everyone from ancient Romans to American Indians. The sonnet includes these lines:

> The strawberry blooms upon its lowly bed,
> Plant of my native soil! – the lime may fling
> More potent fragrance on the zephyr's wing;
> The milky cocoa richer juices shed;
> The white guava lovelier blossoms spread;
> But not like thee to fond remembrance bring
> The vanished hours of life's enchanting spring,
> Short calendar of joys for ever fled!

For centuries, this 'short calendar of joys' has given the strawberry much of its appeal. Its long absence from the table makes the season's first fruit taste even sweeter and more magical. Perhaps because their traditional growing season is so tantalisingly short (a matter of weeks), the British have long been strawberry lovers. In Shakespeare's *Richard III*, the Duke of Gloucester says to the Bishop of Ely: 'When I was last in Holborn, I saw good strawberries in your garden there. I do beseech you send for some of them.' Wimbledon tennis fans consume the fruit with gusto. More than 60,000 pounds of English strawberries are sold during the annual two-week championship.

The French, too, were early admirers of the berry. In 1368, Charles V commanded Jean Dudoy, his chief gardener, to plant 1,200 strawberries in the royal gardens of the Louvre. Five centuries later, Louis XIV launched a literary competition in honour of the strawberry, his favourite fruit. Madame Tallien was perhaps most extravagant in her indulgence. The French beauty, a prominent member of Napoleon's court, is said to have bathed in the juice of strawberries. Even she would be impressed, however, to see the strawberry celebrated so fervently during an annual festival in the small French town of Beaulieu-sur-Dordogne. At the 'Fête de la Fraise', bright red banners depicting strawberries line the medieval streets. Everything from brass bands to ethnic dance groups honour the fruit. Strawberry sellers are, of course, out in force, and the festival culminates in the creation of an immense strawberry tart. Up to 20 feet in diameter (one year it was shaped like a boat), it is displayed in the town centre before being devoured by the crowds.

Their enthusiasm is understandable. There is little to beat a ripe strawberry – lush and red, soft and succulent. Delicately perfumed with the scent of woodlands, it oozes with summer juices. Such a fruit is hard to find today. More than any other fruit or vegetable, the modern strawberry represents an era in which the seasons exert little power over what we eat. Supermarket strawberries have often travelled thousands of miles from fields in California, Mexico or Spain in an Envirotainer (a metal box in which battery-powered fans circulate air over dry ice) inside an aircraft. They are then shunted on to trucks and into supermarket storage units. To minimise damage to the fruit, the food industry favours strawberry breeds such as the Elsanta, a Dutch variety that produces high yields, has a long shelf life and will withstand rough journeys. The Elsanta, however, can also be hard and tasteless – a pale shadow of the fruit that inspired Helen Maria Williams to write her sonnet. With the advent of the long-haul strawberry, we have surely lost the thrill of the seasonal fruit whose first sweet bite confirmed the happy knowledge that summer had finally arrived.

The strawberry is now an avid collector of air miles. If weight-less space travel is a novelty for the fragrant fruit, it is no stranger to jet-powered transport. On airlinemeals.net, a surprisingly entertaining website that posts thousands of photographs taken by travellers of their in-flight dining experiences, the frequent appearance of strawberries in cabin meals can be tracked. There is the photograph taken by Brian on 14 September 2003. On his seven-hour US Airways flight from Pittsburgh to London, his vegetarian dinner included a dish of fresh strawberries. Then on 6 January 2006, the dessert of a meal snapped by John on his business-class Northwest Airlines flight from Detroit to Tokyo consisted of fresh strawberries with chocolate ice cream in a pastry cup. 'The strawberries were fresh and delicious, the ice cream rich and the pastry cup crispy,' John noted approvingly. Travelling economy from London to Colombo on Sri Lankan Airways on 14 September 2003, Siva Ganeshan polished off his Asian vegetarian meal of steamed rice with raisins, dhahi ghobi, tem-pered dhal, spinach and green salad, with fresh strawberries. The rice was 'a bit hard', he wrote, but he deemed the overall meal good 'despite the fact it was served at 23.00 UK time'.

Whether passengers are eating strawberries and cream or chicken in white wine sauce, serving three-course meals in a confined space while travelling at 500 miles an hour at 30,000 feet above ground is not the most sensible of endeavours. As a writer in the *San Francisco Chronicle* once put it, it is rather like 'throwing a dinner party for 200 in three hours using a kitchen the size of an armoire'.

Airline posters from the mid-twentieth century unwittingly highlight the absurdity of high-flying cuisine. In 1958, an Air France advert featured a black and white photograph showing the steps leading up to the door of one of its aircraft. On one side, a smartly dressed air hostess salutes an imaginary passenger. Lined up on the other side is a series of characters – a butler holding a silver-lidded platter, a sommelier proffering a bottle of wine, a chef in white overalls and chef's hat and a peasant woman carrying a basket of fresh vegetables (no doubt some strawberries are in there somewhere). 'Step into France anywhere in the world!' declares the poster's headline. In another ad, a 1966 TWA poster, a lavish array of dishes is displayed on a large table. The spread includes the sort of roasted meats, cheeses, piles of fruit and elaborately decorated desserts sometimes seen in old master paintings. 'There hasn't been anything like our Royal Ambassador first class menu since Henry VIII invented banquets,' the poster boasts.

These days, an airline meal could hardly be further removed from a banquet, even in the first class cabin. On the other hand, given the constraints faced by airline chefs and caterers, it is remarkable that anything remotely edible makes it on to the plane at all. For a start, producing meals that look good crammed into a dish the size of an ashtray with tin foil on top is tricky. Moreover, high altitudes alter the taste buds. Something that might be quite decent on the ground can become unpalatable in the air. Andy Sparrow knows all about this. He has worked in the airline catering business for more than three decades. He first encountered this curious phenomenon more than twelve years ago when working as a British Airways wine buyer. 'We noticed a definite difference between wines tasted at ground level and how they would taste in the aircraft,' says Sparrow. On investigation, the airline found that it was not the wine that altered but the palate of the person drinking it. This was due to conditions on an aircraft that do not exist at ground level – everything from changes in the body's biorhythms to stress patterns, dehydration and the aircraft's vibration. 'We found that more powerful, new world, fruit-driven styles worked very well,' explains Sparrow. 'And if a wine had acidity at ground level, sometimes it could be overaggressive up in the air – which is why you shouldn't see wines like Muscadet on board.'

Jet plane travel dulls the palate more generally, too. Anyone who has found themselves breaking out of their normal drinking habits on board to order a tomato juice or a Bloody Mary – a stronger tasting drink than most, particularly when salt, pepper and Worcester sauce are added – will have experienced this. When Sparrow was working as a flight attendant, he and his colleagues would refer to this as 'the tomato juice syndrome'. All it took, he explains, was one passenger ordering a tomato juice or a Bloody Mary during the bar round 'and you could guarantee that throughout the cabin people who don't normally drink tomato juice would be drinking it'.

Such challenges are nothing compared to the task of making sure the right meal gets loaded on to the right aircraft. An average jumbo jet is loaded with tens of thousands of items, many of them related to the food, and managing them all requires a dazzling demonstration of logistical prowess. Take Continental Airlines. The company produces thirty-three million meals a year and has more than a hundred different menus, which are changed periodically. Each year, the airline boards 55 million sodas and juices, 45 million pounds of ice, 3.5 million quarts of orange juice and 30 million bags of pretzels and peanuts. To reduce errors, Continental has a website used by its catering suppliers. The site provides detailed information on the meals – down to the particular cut of tomato for each class of service – and how they must be boarded on the various aircraft. Every flight has a code indicating what should be put on that particular aircraft. Diagrams of the galley explain that, for example, on a Boeing 777, the economy-class meal goes in position 102 and that plastic glasses should be used. Once the aircraft is in the air, limitations on culinary creativity continue. Airlines must adhere to union regulations giving cabin crew a break between meal servings (this is often why airlines deliver the second meal inconveniently close to landing).

Labour laws were not among the obstacles faced by pioneers of the airline catering industry. In the early days of commercial flight, planes flew beneath the weather, so the more pressing problem was passengers becoming sick during the frequent bouts of turbulence (nurses were often chosen as cabin staff). In the 1920s, weight was also a concern. Wicker chairs were sometimes used as airline seats and, in an effort to reduce its load, Imperial Airways hired fourteen-year-old boys to serve its sandwiches, tea and coffee. The poor wretches were fined if they put on weight – 100 pounds was the limit.

Those were the days when passenger flight was exciting and glamorous. Today, with cost cutting high on the agenda, the airline meal is becoming an endangered species. For the airline industry, in-flight meals cost money. When, in the 1980s, American Airlines was looking to cut its budget, the head of the airline, Robert Crandall, famously came up with the idea of cutting the olives out of dinner salads. This small measure saved the company $40,000 a year. Today, whole meals are being cut from the cabin service. Budget airlines have introduced buy-on-board programmes and while on international routes passengers are still fed, travellers on domestic US flights must often fend for themselves. Food companies, seeing a business opportunity in the new austerity, set up small booths selling sandwiches and snacks in the departure lounge. Hotels have also stepped into the breach, providing picnics for guests to take on to the flight.

Other factors make in-flight food services expensive for airlines. For a start, equipment needs to be installed in the cabin – equipment that takes up valuable aircraft 'real estate' that could be used to cram in a few more passengers. As with ocean-going cargo vessels, every moment that a plane is not in the air, it loses money. Loading the meals on board uses up expensive time on the ground for which airlines pay hefty fees. A late food delivery could hold up a flight, making the airline incur more fees. At the other end, the airline may face penalty charges for not reaching its gate in time, and it may have to wait on the tarmac because it has missed its gate slot. Every tick of the clock costs more money, running into thousands of dollars per minute. No wonder airline companies are cutting out dinner where they can.

Yet food is not going to disappear from aircraft. As passengers tuck into the in-flight meal (at least on those airlines still serving it), beneath the floor in the belly of the plane is a consignment of several tonnes of food – fresh produce from all over the world speeding through the sky. This cargo is part of a growing global movement of goods by plane as supermarkets strive to keep their shelves stocked with fresh fruits and vegetables. As a result, strawberries are available in Britain all year round. Those cartons in the fresh fruit section might look identical from one month to the next, but they will have come from anywhere from Latin America to the Middle East. During the summer season, the UK can supply much of the demand locally. However, even in June, July and August, the US west coast – one of the main sources of strawberries sold in Britain – has to step in to supplement demand. As the British summer comes to an end, shipments from the USA pick up again and, by October, these are joined by strawberries from Mexico and Egypt, as well as by some from South Africa. As one growing season ends, another begins somewhere else. Food is literally flying in the face of nature.

At a dinner party in 1969, Arthur Veysey, the *Chicago Tribune*'s London bureau chief, noticed how 'Jet planes spread the seasonal joys' to British consumers. 'It was an amazing meal,' he wrote in his article:

> First there was a choice of melon or avocado from Israel, shrimp from South Africa, or rock oysters from New Zealand. With the Scotch beef came French beans from east Africa, asparagus from Florida, and a salad of lettuce and tomatoes from California. The dinner ended with a choice of strawberries from Mexico or mangoes from India, and California celery accompanied the English cheddar cheese.

What was worth commentary in 1969 is now routine. To get some idea of the geographic diversity of British food, stroll down the fresh produce section of Tesco, the UK supermarket chain, which has a policy of labelling its products with their country of origin. The company even puts the national flag on the packaging to denote British products. On an early morning in July 2006, plenty of products with Tesco's cheery Union Jack on the packets can be found in the aisles. It is summer in Britain so local strawberries are there, as well as the more unlikely British-grown product, shiitake mushrooms (the label reads 'ideal for oriental cooking'). But alongside the British fare is an astonishing array of global fruits and vegetables. These include grapes from Egypt and Spain, kiwis from Chile, ginger from China (ideal with those shiitake mushrooms), cherries from Turkey, red cabbage, organic on-the-vine tomatoes, closed cup mushrooms and cauliflower from the Netherlands, sweet potatoes from the USA, baby baking potatoes from Egypt, butternut squash from Greece, white garlic, peppers and plums from Spain, green chillies from Zimbabwe, pineapples from Costa Rica, baby clementines, Fair Trade oranges and Golden Delicious apples from South Africa, Braeburn apples from New Zealand, asparagus from Peru, extra fine beans and tenderstem broccoli from Kenya and mangetout from Zambia.

While most of these products travelled by sea or truck, some of this astonishingly global array of food has been flown in by plane. Airlines once treated the cargo business as a means of filling up leftover space after securing the real source of revenue – the passengers. These days, however, the chilled fresh produce packed underneath the aisles on the lower deck constitutes a growing proportion of airline profits. Much of this produce is flown by dedicated freighter airlines with names like Arrow Cargo, Benair and Polar Air Cargo. However, the big passenger airlines are competing with these carriers for a share of the world's 'belly-hold freight'. Frankfurt airport now handles so much fish (about 30,000 tonnes a year) that Germany's largest 'port' for fresh fish is no longer a coastal harbour but a 97,000-square-foot set of cool rooms at an airport where workers in warm jackets shift around palettes, boxes and cartons of chilled goods.

At Heathrow airport, another ambitious centre handles much of the fresh food entering the UK. It is part of a vast complex known as Ascentis, a futuristic structure that is the main cargo hub of British Airways. Compared to this giant unit of industrial infrastructure, the Perishables Handling Centre is not much to look at. The two-storey grey building with a small car park in front of it is undistinguished architecturally. However, this building processes about 85,000 tonnes of perishable cargo a year (including up to eight million punnets of soft fruit) on floor space of nearly 70,000 square feet. On a particularly busy December day in 2000, 460 tonnes of fruit, vegetables and cut flowers passed through its frosty spaces – and among the most popular items in this massive haul were strawberries.

Inside the centre is another world, and it is a cool one. Outside, the British might be battling a summer heat wave but the temperature in the centre is chilly, thanks to giant units whirring noisily on the ceiling. Bright red forklifts buzz around the place. Workers dressed in thick fluorescent jackets and woolly hats keep themselves busy. Standing still for too long at temperatures as low as 2° Celsius is not advisable. In any case, no one can waste a second in getting the goods processed, labelled and packed into a refrigerated truck heading for the supermarket.

The journey of a consignment of Californian strawberries destined for a supermarket in Leeds might look like this: on Monday morning, the strawberries are harvested from a field in Watsonville, one of California's biggest strawberry-growing centres. There, they are packed into boxes and stuffed into a truck heading for the airport to catch that night's flight. As soon as the plane touches Heathrow's tarmac on Tuesday, it is opened up and, from within the cavernous interior, anything from Envirotainers to large boxes and occasionally baskets slides out. Among the haul are the strawberries. They move swiftly out of the aircraft's belly on to roller beds, which trundle them into the back of refrigerated vehicles. These four-wheeled fridges speed off to the perishables centre and offload their contents at one of thirteen large doors (queuing up to unload is not an option here).

At the centre, bulk shipments of produce are 'stripped down' and broken up into individual orders on 'break spurs', the goods being sorted by product type or according to the regional distribution centre for which they are destined. The boxes of strawberries are given branded supermarket stickers. Machines at the centre print information about the product's origin, its weight, sell-by date and price on to blank spaces on the label. This means supermarkets can change their prices at the last minute. Suppose Britain's summer turns, as it often does, to soggy rain. Those strawberries – ordered last week when the heat wave was at its peak – might have fewer buyers than retail planners anticipated. So as the strawberries head towards the UK, sales managers can advise the perishables handling centre to knock a few pence off, or stick an extra 'two for one' label on the box.

Getting a label stuck on delicate fruits covered by only a thin film of plastic is a tricky business. Air pressure provides the ingenious solution. As the punnets of fruit encased in their cellophane bags pass below the labelling machine, the device peels off a sticker and, with a powerful jet of air and a hissing like the sound of a car tyre being filled, the machine shoots the label down on to the cellophane. By the afternoon, the strawberries are labelled, sorted into different orders and packed into a truck. The truck will drive overnight to Leeds. By Wednesday morning the strawberries are sitting on the supermarket's shelves – a couple of days after being picked from their field in California. If all has gone according to plan, every step of their jet-setting journey will have taken place in chilled temperatures.

'That's the cool chain unbroken,' says Gerry Mundy, with a note of satisfaction in his voice. Mundy, the son of a farmer, was a food logistics manager before he joined British Airways. He is now head of what BA calls its 'global perishables' business. 'It's the speed of pullback from the aircraft to those points that's critical. We can do that from wheels down to in the centre in less than two hours, and it's often as little as seventy minutes,' he says. 'That might not sound much but I don't know any carrier in the industry that can match that.' Time is money for Mundy and his customers – supermarkets and other retailers. For these companies, shelf life is everything, and it starts the moment a fruit or vegetable is picked. From then on, every minute spent in ambient temperature contributes to its deterioration.

Speed comes at a price. Shipping goods by air is far more expensive than in refrigerated reefer containers on seagoing vessels. So, given that the reefer is becoming better at preserving fresh produce, why is the food industry turning to jet planes? Part of the answer lies in the growing range of delicate produce now in demand – and for which high prices can be charged – such as tree-ripened peaches or avocados. Moreover, freight carried by aircraft fills unforeseen gaps between the growing seasons of different countries. An unexpected cold snap now means a spurt of extra business for the aviation industry.

In the early spring of 2005, a severe frost hit Spain. The country was unable to supply UK supermarkets with salad vegetables. Prices for lettuce shot up by about eighty per cent and crop damage meant Spanish exporters could not meet their lettuce orders. Suddenly, food that would normally have been trucked across Europe had to be brought by plane to the UK from the west coast of America. This is the sort of thing that gets trade publications excited. 'Carriers muck in to avert "lettuce crisis"', was the headline greeting readers of *International Freighting Weekly* on 21 March that year. 'British Airways World Cargo and American Airlines Cargo have reported unprecedented volumes of lettuce imports from the US, particularly California, over the last few weeks,' wrote the paper's correspondent Will Waters, 'while exports of salad vegetables from Europe have shrivelled to almost nothing.'

Airlines found themselves the unexpected beneficiaries of a lettuce import boom. Gerry Mundy remembers when the 'lettuce crisis' hit. BA was flying in 200 tonnes of salad a week from the USA when normally it would have carried none. 'Absolutely every scrap of space that was available out of the west coast was filled with salad goods,' he says. In short, as our food supply becomes more interconnected, the weather in a far-off country is of acute concern to supermarket managers and freight forwarders. In the world of 'global perishables', a frost in one country can cause jet planes to be stuffed with lettuce in another.

As well as maintaining a ceaseless flow of fruits and vegetables, airlines have found a new source of revenue in an increasingly popular form of food. It is called 'fresh cut', and it is the new frontier for food marketers. Executives talk of 'value-added products' and 'low shrinkage rates' but behind the corporate jargon are small sealed plastic containers of sliced or diced pieces of fruits and vegetables, peeled, trimmed and ready to eat. First, there was lettuce in a bag, washed and cut. Now there are pineapple chunks and slices of apple, or mixtures of vegetables sold as 'Asian Stir-fry Mix' – all capitalising on shoppers' demand for convenience. And if twenty per cent of the produce is lost anyway through cutting and trimming, why pay to ship that extra weight?

Of course, there are still people for whom chopping, slicing and mincing are all part of the pleasure of cooking and eating. Nevertheless, more than enough lazy chefs exist for food suppliers to justify the extra cost of this new form of food delivery. And the time it takes to cook a meal is shrinking. Today, the preferred preparation time is about half an hour. By 2030, some have estimated that the ideal time will be five to fifteen minutes. With cut and peeled vegetables and fruit available, this certainly looks possible.

Keeping 'fresh cut' fresh is a challenge, since fruits such as apples and pears turn brown soon after being sliced. Modified atmosphere packaging (MAP), often used in ready-cut bagged salads, introduces nitrogen or carbon dioxide into a container to reduce the amount of oxygen inside it, slowing the growth of unwanted life forms and preventing browning and wilting. Strawberries inside MAP packaging keep for up to two weeks after being picked. New ideas are being explored. One US scientist, Olusola Lamikanra, has been fooling the sensory system of plants. He has found that cutting fruits and vegetables while submerged in water creates a barrier, preventing the onset of the physiological and biochemical changes that occur when a plant is injured – changes that trigger the fruit's deterioration. Breathable plastic films are being developed, using laser scoring systems to tailor levels of permeability to suit the requirements of different foods. A packet of pineapple can, for example, explode if too much carbon dioxide builds up inside it.

Of course, whether or not this packaging is healthy or preserves taste and quality is another matter. There is evidence that modified atmosphere packaging destroys vital nutrients. In any case, fruits and vegetables that have come out of plastic packets tend to be far less delicious than those picked and eaten the same day. So an incident that took place in Britain in May 2006 had some scratching their heads. Sixty-nine-year-old Val Salisbury was found uprooting strawberry plants from a farm near her home in Herefordshire. This act of vandalism seemed odd given that countryside communities have waged a lengthy battle to persuade supermarkets to buy more British food.

In the 1990s, the British strawberry was becoming increasingly rare. Between 1985 and 1995, the percentage of home-grown strawberries sold in UK shops fell from seventy per cent to fifty-one per cent – yet Britain's appetite for the berry was increasing. Suppliers in Spain and California, it turned out, were meeting demand. In 1995, during the British summer, more than 1,000 tonnes of Californian strawberries arrived in the UK by plane and more than 12,000 tonnes of Spanish strawberries appeared in the shops. Environmental groups worried about the emissions created by hauling fruits such long distances. Local farmers complained of falling revenues. Consumers condemned the foreign imports as tasteless fruits with an unappealing texture. 'It will not be long before watery foreign imports outnumber the indigenous,' lamented the *Independent* in a 1997 article headlined 'In memoriam – the British strawberry'.

Then UK farmers started using polytunnels – semi-circular temporary greenhouses made from steel hoops covered in polythene sheeting. These tunnels not only extended the UK's strawberry growing season from about six weeks to six months. They also allowed farmers to cut back on their use of pesticides. Moreover, the tunnels protected the fruit from the sudden downpours that can turn fields of ripe strawberries into piles of pink mush. British farmers were now able to beat off the foreign competition and restore the British strawberry to British shoppers – at least for a larger portion of the year.

All this was lost on some. People living near the polytunnels said they marred the landscape ('They are making such a mess of this beautiful countryside with all of these tunnels and there seems to be no end to them,' Val Salisbury told reporters). Their plastic sheeting, complained rural residents, had a tendency to rattle in the wind and, in the sun, generated a blinding glare as light bounced off shiny plastic. Others said they caused flooding. Homeowners even accused the tunnels of dampening property values. Shortly after Salisbury made her feelings known to her local strawberry farm, Bill Wiggin, the MP for Leominster, a small Herefordshire town, and agriculture spokesman for the opposition Conservative Party, weighed in against the British strawberry. 'I have not eaten a strawberry for about four years now and, as far as I'm concerned, anyone eating strawberries out of season is encouraging the blight of polytunnels on the landscape,' he told *The Times* in June 2006, a comment that enraged farmers.

As this eccentric battle illustrates, efforts to make our food supply more sustainable run up against all kinds of conflicting interests. Yet at a time when many people are starting to argue for growing food locally, the opponents of polytunnels, which gave the British strawberry a new lease of life, seem to be swimming against the tide. In any case, the tunnels (the look of which could easily be improved) are simply the latest addition to a landscape that has always been shaped by agriculture. The rolling hills of England may look the picture of natural beauty, yet there is hardly a square foot of the country that has not over the centuries been carved up, grazed, planted and replanted by those in the business of food production.

When it comes to jet-setting food, however, there are more important tradeoffs to consider than whether or not plastic tunnels are spoiling the view. One of the worries is what aircraft are doing to the environment. Aviation is thought to generate only about three per cent of the world's man-made greenhouse gases and aircraft are far more fuel-efficient than they were even a few years ago. However, air travel is growing fast and aircraft burn high-quality kerosene jet fuel that, on international flights, is artificially cheap compared to other fuels because it is not subject to tax. Moreover, airfreight emits substantially larger amounts of carbon dioxide per unit of cargo than if it were moved by ocean. Concerns focus not only on carbon emissions but also on other substances that the aircraft leave behind them – contrails, those white streaks painted across the sky in the wake of a plane, created by the condensation of water vapour in the plane's exhaust. Some believe contrails, which are released at a high altitude, may have a more potent effect on the environment than gases released on the ground. The trouble is that while in other forms of transport technological fixes are emerging, those solutions cannot be used in aviation. Ethanol is not dense enough to replace kerosene and biodiesel could clog the engines of planes.

One possibility is to use container ships rather than planes wherever feasible. However, for highly perishable produce, aviation is the only answer. As ethical consumers push companies to cut their most damaging food miles, some retailers are even letting their customers make the choice, adding a label to produce that has travelled by air. While Tesco embarks on its project to estimate the overall carbon footprint of the things it sells, the company has instigated an interim measure – putting an airplane symbol on produce imported by air. Marks & Spencer is doing the same. Its air-freighted products will get a 'flown' sticker on them.

Yet jet-setting fodder is also responsible for developments that are more positive. For poor countries, the ability to sell 'value added' vegetables to western markets generates badly needed jobs for farmers and other workers that might not exist, were we to return to buying everything locally. Local food proponents like to point out that they are supporting local farmers, but there may be 'local farmers' elsewhere who could also benefit from selling their produce. 'The San Francisco Bay region is one of the wealthiest local economies on the entire planet,' write the ethicists Peter Singer and Jim Mason. 'If we have the choice of using our purchasing power in our local economy, or buying products imported, under fair terms of trade, from some of the world's poorer nations, is there any merit in keeping our money within our own community?'

USAID, the development agency, would agree. In Ghana, USAID has worked with the Dutch food retailer Royal Ahold to help local companies equip themselves to produce pre-cut pineapples and fruit salads that meet US and European quality and health standards. The organic pineapples of Ghanaian farmers are now fetching high prices, raising thousands of dollars a year in community development funds. Some think the growing demand for organic food in the USA and Europe could give farmers in developing countries a chance to boost their incomes. Moreover, organic farming techniques tend to be environmentally sustainable and suitable for smallholder farmers. In Uganda and Tanzania, farmers involved in a project to promote organic exports run by Sida, a Swedish government development agency, reported that they had increased their harvests by up to one hundred per cent and were receiving between twenty and fifty per cent more money for their organic produce.

Even in non-organic farming, noticeable improvements in the lives of people in developing countries lie behind the figures. In Kenya, where horticulture is the fastest-growing sector of the economy (making up more than twenty per cent of its agricultural exports), the industry not only provides much-needed work for unskilled labourers in rural areas where there are no other job opportunities, but also purchases produce from smallholders, generating income for them. As a result, the industry has raised living standards – particularly for smallholders – according to the University of Sussex. In rural areas, it found that households of Kenyan horticultural smallholder farmers more often had access to piped water than other families. Eighty per cent of horticultural smallholders had concrete floors in their homes, compared with thirty-five per cent for those not involved in the horticultural industry. In urban areas, the researchers found that, while wages remained low, there was less evidence of food poverty among those working in the horticultural industry, which also provided jobs for unmarried women with no other employment prospects.

The importance of this kind of trade to African countries – and its dependence on transport – was revealed in 2005, when

British Airways suspended its weekly flights from Lusaka to Spain, giving as the reason the cost of fuel. A large farm in Zambia that employs about 3,500 people said the flight cancellations would cost the business (which supplies vegetables and cut flowers, mainly to British supermarkets) almost $100,000 a week. Exporting fresh vegetables and cut flowers to Europe, the chief executive of the Zambia Export Growers Association told *The Post* of Zambia at the time, brought the country revenues of more than $60 million a year – valuable income that would be lost without the assistance of the jet plane.

Britain's consumption of fresh fruits and vegetables supports the livelihoods of more than one million Africans, according to the London-based International Institute for Environment and Development. The institute claims that transport of this produce by plane from Africa to the UK accounts for less than 0.1 per cent of the UK's total carbon emissions. 'In the big picture, the environmental cost of international food transport is trivial compared with UK domestic food-miles. Plus, air-freight is the only possible mode of transport for some highly perishable produce where no other infrastructure exists,' write James MacGregor and Bill Vorley in a 2006 paper published by the institute. 'It is time to look to the huge impacts of the food system at home, rather than pull up the drawbridge on Africa.'

Poverty is not bad only for humans. Development experts argue that poverty is also environmentally damaging. People without jobs or access to other income are more likely to turn to the natural resources around them, felling their forests for the wood or emptying their lakes and rivers of fish, reducing biodiversity and creating long-term environmental degradation, air and water pollution. Of course, industrialised agriculture can be damaging, too. Large farms in places such as Kenya have been accused of draining local resources such as water. Agribusinesses can create pollution by using chemical fertilisers. However, if it is properly managed, farming is a more sustainable option (for developing economies as well as for the environment) than over-fishing or forest felling. So when we accuse the airborne South African strawberry or the high-flying Kenyan bean of destroying the planet, are we really picking our most pernicious adversaries?

11

Elevated Design

Buffalo grain feeds the Bauhaus inspiration

Wheat flour: Powdery foodstuff obtained by grinding the grain of a cereal grass into meal

Origin: First domesticated in the area known as the Fertile Crescent, which today covers parts of Jordan, Iraq, Syria and Turkey

Etymology: Middle English *whete*, from Old English *hwaete*

Legends: On 13 December, Santa Lucia's day, Sicilians will not touch pasta, bread or anything made of wheat flour but eat *cuccia* (boiled whole wheat, ricotta and sugar), a dish associated with one of the legends surrounding the saint. When famine struck Syracuse in Sicily, Lucia sent its people ships loaded with wheat. The hungry citizens did not wait to grind the grain but boiled it instead

*I*N THE 'POINTS of Interest' section of *New York: A Guide to the Empire State*, a 1940s guidebook, the authors describe the harbour at Buffalo, New York. They highlight the frenzy of activity taking place along the waterfront that stretches for 37 miles east and west from the south end of Michigan Avenue. The scene, we are told, presents 'a superb panorama of commercial and industrial Buffalo, with smoke-stacks of the Lackawanna plant of the Bethlehem Steel Company to the left, and the elephantine grain elevators, grain-dusted in mist or sun, to the right'. The authors convey the impression of a city alive with the constant flow of activity. Water-borne transport vessels busy themselves round the harbour and giant soot-blackened machines clank and spit. 'Noisy switch-engines back produce-laden freight cars to the doors of grim warehouses that line the many slips,' the guide-book continues. 'Fussy tugs churn the harbor waters and dash spray over the breakwall from rough Lake Erie beyond, where the Coast Guard station, white and neat, sits at the foot of the old lighthouse with its mooing foghorn.'

This industrial hubbub has long since disappeared. Today, the waters of the Buffalo River are calm and oily. A handful of vessels ply the waterfront – some of them tourist boats. Buffalo's economy went into decline in the 1950s and by the 1970s residents were leaving the city in droves. The population has shrunk from more than half a million in the late 1950s to 290,000 today. City officials and business leaders are exploring Buffalo's potential to become a biotech centre. But bio-technology is an invisible industry, hidden within the walls of pristine labs and squeaky-clean processing facilities. None of the noise, smoke and commotion that inspired the guidebook authors will ever return, even if Buffalo's fortunes do.

Along the waterfront, however, there is a lasting reminder of the source of Buffalo's once mighty economic position – the 'elephantine grain elevators' described by the *Guide to the Empire State*. With names like Great Northern and Concrete Central, these were the machines that moved the powerful grain industry, transferring cargo mechanically between ships and canal boats at great speed. In shifting mountains of grain, these lumbering machines turned nineteenth-century Buffalo into the world's largest grain supplier and one of the wealthiest towns in America.

Many of the city's grain elevators still stand, abandoned and dilapidated, clustering round the water's edge, the forlorn relics of a once booming business. Their vast concrete walls are stained and broken glass fills the windows that puncture their giant towers. Saving these monumental structures is the mission of a group of Buffalo-based academics and architects. Two of them (the buildings, not the academics) have been listed on the New York State and National Registers of Historic Places, although, sadly, one of the two – the last surviving wooden elevator – was recently lost in a fire. Photographers have documented them in exhibitions and books. Boat rides now take tourists and architecture fans along the Buffalo River to view them. Local artists have been depicting them in oils and watercolours. So what makes these crumbling industrial relics so special?

For a start, they are huge. Looming over the water, these monolithic, grey clusters of cylinders and towers are wildly impressive; monsters of mankind's making. Some look like medieval fortresses or prison blocks. Others seem to be closer in spirit to ancient religious structures such as Stonehenge or the Easter Island statues. When British novelist Anthony Trollope visited Buffalo in 1861, he compared them to dinosaurs. 'An elevator is as ugly a monster as has been yet produced,' he wrote in his travelogue, *North America*. 'In uncouthness of form it outdoes those obsolete old brutes who used to roam about the semi-aqueous world, and live a most uncomfortable life with their great hungering stomachs and huge unsatisfied maws.'

A quarter of a century later, a group of Europeans saw them in a different light. In 1913, an article appeared in which Walter Gropius, the German architect who was one of the founders of the Bauhaus school of design, published a series of photographs of American elevators, storage silos and factories. The images prompted Erich Mendelsohn, a prolific modernist architect, to visit Buffalo in 1924. 'I took photographs like mad,' he wrote to his wife. 'Everything else so far seemed to have been shaped interim to my silo dreams. Everything else was merely a beginning.' As photographs of the elevators appeared in European publications, other architects and designers grew excited by the size and form of these unfamiliar structures and started incorporating radical new ideas into their work. As they did so, the significance of these industrial machines extended beyond the movement of food. They became a source of inspiration for modern architecture.

Joseph Dart, a nineteenth-century entrepreneur, and his engineer Robert Dunbar, a Scotsman, were not contemplating art when they devised the giant machines that would become the engines powering Buffalo's prosperity. Today, Dart and Dunbar lie buried in the city's green and tranquil Forest Lawn Cemetery surrounded by classical columns and Roman-style sarcophagi, but during their lives, they immersed themselves in the world of commerce and technology, enthusiastically embracing the instruments of the modern industrial age. Born in Connecticut, Dart arrived in Buffalo in 1821 and soon established himself in the hat and fur trade. This did not last long. Dart, who had the entrepreneur's knack for spotting potential business opportunities, quickly recognised what the real source of wealth in Buffalo would be: wheat.

In America, this nutritious grain had risen at the hands of colonial farmers to a position as the USA's pre-eminent crop. Even before the American Revolution, a casual visitor to what were then British colonies might have been able to predict the rise of US wheat to global dominance. Field upon field of the golden crop, waving and shimmering in the breeze, dominated the landscape, extending north into New York's Hudson Valley and west into the Mohawk Valley. By the late eighteenth century, farmers were no longer modest settlers eking out a subsistence living from the land. They were skilled agriculturalists seeking new markets. Moreover, with war consuming Europe's energies, demand for American wheat was growing rapidly. Between 1839 and 1859, American wheat production more than doubled, as it did during the boom years after the Civil War drew to a close in 1865. While, between 1775 and 1783, the American War of Independence created a temporary setback to commercial cultivation, the USA was in the nineteenth century becoming a powerful player in world grain markets.

Wheat would transform Buffalo's fortunes. Sited on the eastern shore of Lake Erie, what started life as a sleepy village called New Amsterdam was, by the 1890s, a wealthy metropolis and home to sixty millionaires, more than any other

US city at the time. Streets were being paved. Gas lighting was being installed. Sewers were being laid. Soon Buffalo had an impressive array of civic buildings, ostentatious hotels, capacious churches and elegant parks. Neo-classical style banks welcomed new customers and department stores offered the latest fashions. Grand residences emerged all over the city, such as the Mansion on Delaware Avenue, built in 1867 by architect George Allison (the house is now a boutique hotel). Hydroelectric power generated by the Niagara Falls made Buffalo the first American city to have widespread electric lighting. It would be home to two US presidents, Millard Fillmore and Grover Cleveland (it would also be the scene for a presidential tragedy when, at the 1901 Pan-American Exposition, William McKinley was assassinated).

It does not take long to grasp the fact that Buffalo was once a place of some substance. In a plan that echoes the grandiose symmetry of Paris or Washington DC, broad avenues fan out from Niagara Square, which is overlooked by the gargantuan art deco City Hall (the tallest in the USA). Frederick Law Olmstead, the man behind New York's Central Park, created an ambitious series of parks and interconnected parkways for the city. Charles Clark, the Buffalo lawyer who established the cemetery where Dart and Dunbar now lie, took his inspiration from Paris's famous Père Lachaise. Over on Main Street, the *beaux arts*-style Buffalo Savings Bank, built in 1901, looks more like an opera house than a financial institution. It is topped with a large gilded dome (the last restoration used 140,000 sheets of gold leaf), lavished with columns and architraves and, inside, its walls are covered in elaborate murals. Nearby, the massive Italian Renaissance-style Ellicott Square Building was the largest office building in the world for sixteen years after it opened in 1896, with a mosaic in its interior courtyard made up of twenty-three million pieces of marble imported from Italy. The aspirations were global, the styles ostentatious. Today, Buffalo might be an under-populated city at a low economic ebb, but in the nineteenth century it was clearly the place to be.

Buffalo's affluence was built on grain. However, the city's entrepreneurs were not farmers. They were engaged in the business of trading the grain that farmers produced. Meanwhile, the source of grain was changing as the 'bread basket of America' shifted from the eastern part of the country to the Midwest. As hundreds of thousands of people moved west, a prosperous farming economy emerged on the fertile plains of states such as Ohio, Illinois, Wisconsin and Indiana. Wheat from those plains was moved from one side of the country to the other, and on to far-off markets in Europe.

What made this possible was the creation of a vital link in the country's transport network – a link that was established shortly after Dart arrived in Buffalo. In 1825, the Erie Canal opened. To celebrate the event, New York State Governor DeWitt Clinton – a savvy operator who had thrown all his political energies behind the canal project – travelled eastwards from Lake Erie to the canal's terminus at the Hudson River near Albany, more than 300 miles away. He then continued down the Hudson to New York Harbour, where he emptied casks of water from Lake Erie into the Atlantic Ocean. It was a momentous occasion. The canal was one of nineteenth-century America's most important public works, creating a vital link between Midwest America and the eastern seaboard. As well as accelerating the development of the Midwest, it transformed the young country, sparking a wave of capital investment and innovation. It created a watery interstate network along which the country could connect itself to itself – and then to the world beyond its shores.

The opening of the canal had a dramatic effect on the movement of grain. Before Governor DeWitt ceremoniously poured his Lake Erie water into the Atlantic, long and arduous routes across the USA had been necessary. Even grain from western New York destined for New York City was sent in barges down the Ohio and Mississippi Rivers to New Orleans to be put on a ship, which would then sail back north up the east coast – a journey of 3,000 miles – because this was cheaper than sending it overland. The Erie Canal eliminated the need for these long and difficult routes. Freight charges fell from $100 for a tonne of grain for overland transport to just $10. Freighter vessels could now ship grain across the Great Lakes to Buffalo, from where it continued its journey east to New York and then on to Europe. Sitting between the easternmost point of the Great Lakes and the start of the Erie Canal, Buffalo became a critical connection point at the heart of trade that stretched from Midwest America to Liverpool. For the grain industry, then, Buffalo was rather like Memphis is for FedEx packages – a massive transit point, the central link in the global grain chain.

The implications were not lost on Dart. He watched the rapidly rising grain receipts with intense interest. In a paper read before the Buffalo Historical Society in 1865, he wrote:

> It began to be evident that there was to be a very speedy and immense increase in the future grain business of this port. It seemed to me, as I reflected on the amazing extent of the grain producing regions of the Prairie West, and the favorable position of Buffalo for receiving their products, that the eastward movements of grain through this port

would soon exceed anything the boldest imagination had conceived.

Dart was right. In 1836, Ohio's grain crop for the first time surpassed that of New York – and 1.2 million bushels of it passed through Buffalo. By 1846, more grain was being shipped through Buffalo than Ohio. In 1855, the Board of Trade and Commerce declared: 'Buffalo is now universally acknowledged to be the greatest grain market on the Continent, not even excepting the City of New York.'

Buffalo, now known as the 'Queen City of the Lakes', was at the heart of an agricultural industry of immense global importance. Along with cotton, American grain made up by far the largest proportion of imports going to Britain. In 1846, the British repeal of the Corn Laws that had limited grain imports led to a rapid increase in the sale of Midwestern wheat to the country. And it was not just Britain that felt the effect of the cultivation of wheat on far-off American prairies. During the late 1860s and 1870s, it was American exports that were instrumental in shaping developments in international grain markets.

The problem for grain traders was that because the lake boats could not fit into the canal and canal boats, pulled by horses, could not travel on the lakes, a massive handover operation from one vessel to another had to take place before the grain could continue its journey east. This was a complicated and laborious process. Grain was generally transported loose, rather than in sacks, and handling the stuff was gruelling work that was also tremendously inefficient. As soon as a lake boat had docked, Irish immigrant labourers would descend deep down inside the holds of the lake vessels and, using only shovels, heave thousands of tonnes of grain into wooden barrels, which were then carried by the stevedores up to the warehouses. As more and more grain flowed through Buffalo, this system was clearly going to be unsustainable.

Dart saw a business opportunity. 'In these circumstances, I determined, in 1841, to try steam power in the transfer of grain for commercial purposes,' he later wrote. He had seen steam power at work in another context. In the 1780s, Oliver Evans, a Delaware-born millwright and inventor, had developed a steam-driven system of buckets on a revolving chain – a sort of conveyor belt – to raise grain up and use the force of gravity to move it though his flourmills. Dart reckoned he could adapt this technique to the transfer of grain from one vessel to another. In 1842, building on Evans's design, Dart and Dunbar devised a steam-driven belt running up a wooden leg (known as a 'marine leg') that, like the proboscis of a giant mosquito, could be lowered into the ship's hold. Along the belt was a series of buckets – initially 28 inches apart and holding about 2 quarts each – that would mechanically scoop up the grain, lift it into a wooden structure, weigh it and drop it into tall storage bins. Instead of heaving the grain out of boats on their backs, workers would stand in the vessel and shovel it on to the moving buckets.

Not everyone was convinced by the idea. When Mahlon Kingman, a prominent Buffalo forwarding agent, learnt of the machine, he told Dart: 'I am sorry for you; I have been through that mill; it won't do; remember what I say; Irishmen's backs are the cheapest Elevators ever built.' Kingman was later forced to eat his words. When the elevator was up and running, and one afternoon he came to Dart to get two of his canal boats loaded, he sheepishly confessed: 'Dart, I find I did not know it all.'

The machine's capabilities were impressive. It could unload, raise and weigh at least a thousand bushels an hour, a process that a decade earlier had taken several hundred workers several days. Rudyard Kipling, who visited Buffalo in the 1880s, described how 'the glittering, steel-shod nose of that trunk burrowed into the wheat, and the wheat quivered and sunk upon the instant as water sinks when the siphon sucks, because the steel buckets within the trunk were flying upon their endless round, carrying away each its appointed morsel of wheat'.

The elevator's ability to dispatch the grain into waiting canal boats was equally efficient – once it had been lifted into the bins, gravity did the rest of the work. Opening a small trap at the base released the grain, which poured out into the boats below the dock. Even in storage, the commodity was frequently on the move since, when kept in great concentrations, grain becomes highly combustible. Every couple of days, therefore, the grain was lowered and raised into fresh silos in order to keep it cool. 'And thus rivers of corn are running through these buildings night and day,' wrote Trollope in awe. 'The secret of all the motion and arrangement consists, of course, in the elevation. The corn is lifted up; and when lifted up can move itself and arrange itself, and weigh itself, and load itself.'

The new system was a raging success, and by the 1850s, Dart's tall and initially lonely-looking wooden elevator had company. Ten of them – some up to 250 feet high – clustered around Buffalo Harbour, all of them busily shunting grain from ships to canal boats or to storage bins. Construction of elevators continued. When Dart's paper was read to the Buffalo Historical Society in 1865, he wrote,

> There are now twenty-seven Elevators, besides two floating Elevators; storing, some of them, six hundred thousand bushels, and all together fully six million bushels, and capable of moving in a single day, more than the entire annual receipts at this port at the time my Elevator was built ... Not all the product of the gold mines of California will equal the value saved to the internal commerce of the Western and Northwestern states, by those labor-saving Elevators, with only the improvements in them now in use.

The improvements kept on coming. The distance between the buckets on the belt was shortened, speeding up the elevation rate to 7,000 bushels an hour. Dunbar continued to build bigger and better elevators. The larger the elevators, the more could be stored, securing a supply in winter, when the lakes froze, preventing the arrival of vessels. Other entrepreneurs followed, adding functions to Dunbar's design. Horizontal troughs, channels and gutters introduced inside the structures meant grain could be distributed to bins that were not located next to the elevator leg. Some marine legs clung to the sides of the buildings, clad in corrugated iron (they were known as 'stiff legs'). Others had wheels at their base (these were 'loose legs') and moved back and forth along the side of the building on steel tracks. There were even floating elevators that could assist those without their own unloading machines. Electricity generated by the Niagara Falls took over from steam when it came to powering the elevators (also helping Buffalo become a flour-milling centre).

By the mid-1880s, the largest of the elevators were huge vertical warehouses that sucked grain up from one vessel and spat it out into another. They could store a million bushels of grain – enough to make almost seventy-three million loaves of bread – and empty boats at a rate of 19,000 bushels an hour. Their design, too, was shifting away from the original tall, square-sided structure to rows of cylindrical bins (more like their twentieth-century concrete equivalents) with horizontal and vertical 'head-houses' containing the leg and the belt of scoops or buckets. Wood was soon abandoned. Too many elevators had been lost in a burst of flames. Designers experimented with steel bins and ceramic tiles before reinforced concrete became the norm. Even so, explosions – the ultimate disaster for the industry – remained a constant threat, since grain dust is highly flammable.

Reinforced concrete permitted the construction of even more grandiose structures, some the size of city blocks. It is these later versions that survive. From a distance, they give the impression of a small metropolis of art deco buildings. As you draw alongside them, clustering along the banks of the Buffalo River, it is as if a corner of Venice has been recreated in concrete and enlarged to accommodate giants. Close up, fantastic angles and dramatic lines appear, unlike those of any urban setting. While the streets of downtown Buffalo form an elegant symmetry, the elevators were sited at the water's edge and so sit at odd angles to each other as they adhere to the river's bends and twists. Individually, too, these buildings are remarkable structures. Painted in white, the General Mills complex (formerly the Washburn-Crosby and one of the city's few working elevators) is a massive cluster of towers, cylinders and pipes. Passing it, the sickly smell of caramel and oats fills the air as the Cheerios being produced by General Mills inside make their presence felt.

Even without the smell, the elevators are hard to miss. The Great Northern rises up beside the river like a huge fortress. About 300 feet in length and equivalent in height to a ten-storey building, it was designed by Max Toltz, the engineer behind the bridges of the Great Northern Railway Line. It was one of the largest elevators in the country when it was completed in 1897. While it might look like a fortress, its immense red brick curtain walls were not created to prevent the onslaught of an army. The walls were simply protection from the weather for the 99-foot-tall cylindrical steel grain bins they concealed. Later structures abandoned these protective walls, exposing the cylindrical storage bins.

By the twentieth century, the volumes of grain were such that it was these bins, not the elevator machinery, that were the dominant part of the structure. Concrete Central is the queen of these majestic industrial monuments. Built in three sections from 1915 to 1917 by Harry Wait, the man behind many of Buffalo's elevators, she is a quarter of a mile long and 150 feet high and could store 4.5 million bushels of grain. The largest container ships would be dwarfed when drawing up at the wharf beneath her massive façade of ranks of cylindrical towers.

Inside, the elevators echo the ancient churches of Europe, with forests of thick columns terminating in fan vaulting – except that instead of the delicate Gothic ornamentation used by medieval craftsmen, these were fashioned from moulded concrete. British architectural historian Reyner Banham compares the interior of the Great Northern's horizontal head-house to that of a cathedral, 'long, lit by ranks of industrial windows in the corrugated roofing on either side, filled with a golden-gray atmosphere of flying grain dust sliced by low shafts of sunlight'. Some of the elevators have covered walkways round their bases, like tiny cloisters hugging the foundations of a giant monastery. High passageways run between buildings at such strange angles that, outside, the sides of the elevators suddenly appear to slope alarmingly. Even humans feel like a commodity in these buildings, for the quickest method of reaching the top is based on the same principle used to raise the grain: the conveyor belt. With a handle for clasping and a footrest, the 'man lift' is a vertical machine for moving people. It shoots towards the sky at great speed, penetrating storey after storey, carving its way up through the building. Simply step on and off at the floor you need to access.

Trollope was both fascinated and repelled by the brutish efficiency of the grain elevator machines he encountered in 1861. Describing the trunk or proboscis of the elevator as it plunged into the ship's hold, he wrote: 'When there, it goes to work upon its food with a greed and an avidity that is disgusting to a beholder of any taste or imagination.' Trollope added a final disdainful note to his description of the elevators:

> I should have stated that all this wheat which passes through Buffalo comes loose, in bulk. Nothing is known of sacks or bags. To any spectator at Buffalo this becomes immediately a matter of course; but this should be explained, as we in England are not accustomed to see wheat travelling in this open, unguarded, and plebeian manner. Wheat with us is aristocratic, and travels always in its private carriage.

The Buffalo system may have seemed 'plebeian' to the English novelist, but this system had turned the city into the largest grain port in the world.

Day in, day out, grain continued to rush through the elevators. By the 1920s, it passed through Buffalo at a rate of more than 300 million bushels a year – which would make enough bread to feed today's Americans for about two years. While Buffalo businessmen were designing new forms for their elevators (using innovative materials such as concrete and steel to replace highly combustible wood structures), Europe was also focusing on the design of the city's elevators. However, while in Buffalo, the concern was greater efficiency in transporting grain, European architects were using images of the elevators to formulate a new theory of architecture – modernism.

Erich Mendelsohn was not the only architect to be inspired by the 1913 publication by Gropius of photographs of the grain elevators. In 1929, Bruno Taut, a German architect known for his theoretical writings, included a view of Buffalo's Concrete Central in his *Modern Architecture*. Around Europe, the images of elevators appeared in the published works of dozens of designers and architectural theorists. However, it was Le Corbusier, the radical Swiss-born modernist architect, who most passionately promoted American industrial forms as exemplars of a universal architecture. He was extremely excited by the elevators. In his 1923 manifesto, *Towards a New Architecture*, he reproduced images of them and hailed them as 'the magnificent FIRST-FRUITS of the new age', using capital letters to underline the urgency of his message.

Reyner Banham, who worked at the State University of New York in Buffalo in the 1970s, was the first person to examine this

elevated design

powerful connection between American industrial structures and modernist architecture. In *A Concrete Atlantis*, he argues that the daylight factory (a reinforced concrete, steel and glass industrial building) and the grain elevator – seen through the photographs published by Gropius – played a crucial role in shaping modernist ideas. 'Gropius seems to have produced a crucial change in the sensibilities of modern architects, bringing together a set of images whose time had come in the development of European architectural sensibility,' writes Banham. 'The stage was set for the legitimization, so to speak, of industrial forms as the basic vocabulary of modernism.'

Not everyone held these industrial buildings in high regard. Four years after Le Corbusier wrote his manifesto, the *New Yorker* reported that 'we said that a new office building looked like a grain elevator and got sued for $500,000, no less'. But for the early modernists, the *New Yorker*'s comment would have been a compliment. The elevators with their industrial efficiency embodied ideas of progress and the modernists enthusiastically embraced the concept of mass production, pursuing forms that were austere, geometrical and made use of new industrial materials. Le Corbusier, a passionate lover of reinforced concrete, famously saw houses as 'machines for living in', so the monolithic towers and utilitarian box-like structures lining the Buffalo River fitted in perfectly with his vision of the future.

Ironically, the grain elevators were not exactly the structures of the future. As modernist architecture was gaining ground in Europe, the end was not far off for the elevators. In 1932, the opening of the Welland Canal, which ran through Ontario, presented the first challenge to Buffalo's pre-eminence as a transhipment centre, since this waterway, which could accommodate grain boats, bypassed the city. The fate of the elevators was sealed by the opening in 1959 of the St Lawrence Seaway, a joint American–Canadian project creating a series of locks, canals and waterways that connected the Great Lakes to the Atlantic Ocean, rendering the Erie Canal obsolete. Now grain could be loaded directly on to ocean-going ships at ports such as Chicago and Detroit. Buffalo was no longer needed.

In another sense, too, the modernists were mistaken in embracing these industrial buildings as symbols of modern life. While their architecture was moving towards a world where concrete, glass and steel would permit ever more lightweight structures, the grain elevators were heavy, solid and earth-bound. To be sure, they used concrete and steel, the materials of modernism. However, the purpose of the companies building them was not to create airy spaces with a minimum of materials but to construct robust containers capable of holding about 30,000 tonnes of grain. This put a load of about 10,000 pounds per square foot on to the base of the structure, necessitating thick concrete foundations. Pilings had to be driven up to eighty feet down below the surface until they hit solid rock. Great thick concrete walls reinforced with steel were needed to prevent the outward pressure of such a large amount of grain from bursting through its restraints.

Nevertheless, the aesthetic influence of the grain elevators was far reaching. One only has to look at Gropius's 1911 Fagus Shoe Factory or Taut's 1928 Carl Legien Housing Estate in Berlin (where ranks of balconies stacked vertically puncture the façade like large marine legs) to see the visual connections with structures such as Concrete Central. The Bauhaus school building itself, with its simple, solid geometrical form and flat roof, might not have looked out of place along the banks of the

Buffalo River. Le Corbusier's famous (and, thankfully, unrealised) 1925 plan for Paris, in which he proposed flattening great tracts of the city, echoes the grandiose scale and stark simplicity of the grain elevators in the gargantuan towers he envisaged marching across the French capital. Even brutalist architecture of the 1960s such as west London's Trellick Tower – designed by Erno Goldfinger, who knew Mendelsohn – resembles the monumental geometric forms of the elevators.

The elevators provided inspiration for a new language of architecture that derived more from a vague feeling about them than from any technical study or slavish copying. After all, the only European designer who had actually visited Buffalo was Mendelsohn. Le Corbusier had simply reused (and tampered with) the photographs that had been published by Gropius, which were already grainy. What was going on was a game of visual Chinese whispers – but one that would have a powerful impact on the designs of the future. In the hands of Le Corbusier, the elevators had become architectural icons. 'The forms of factories and grain elevators were an available iconography,' writes Banham, 'a language of forms, whereby promises could be made, adherence to the modernist credo could be asserted, and the way pointed to some kind of technological utopia.' The fact that the elevators were an American invention was important, too. Here was the new world offering inspiration to old Europe.

Thousands of years earlier, a far humbler transport device also played a crucial role in the evolution of art and design. In the sixth century BC, Greek artists decorated food and drink vessels with elaborate images of battles and hunts and gods and satyrs. They used a type of slip (a liquid form of clay) that, on firing, turned black and so stood out against the natural terracotta colour of the pot. Using this slip, the artists skilfully traced the forms of human figures, arranging the silhouetted characters in their scenes around a three-dimensional object with rhythm and delicacy. Some identify the black-figure decorations of this early Archaic period as among the first examples of figurative European drawing and painting. While the Greek vessels were not shipped on anything like the scale of their later cousins, the Roman transport amphorae, they were exported widely, spreading their creative spirit throughout the ancient world.

Wooden barrels are also found in art. In fifteenth-century Italy, barrels – by then ubiquitous as transport and storage containers – had a hand in the development of perspective in Italian Renaissance painting. Art historian Michael Baxandall argues that trade and transport were such an all-pervasive part of life that most people were used to calculating the volume and weight of barrels, which at the time did not come in standard sizes. The painter Piero della Francesca, Baxandall explains, wrote a handbook for merchants, *De Abaco*, in which systems for gauging the volume of a barrel are laid out. This book was used in schools, where children learnt mathematics – particularly the sort used by merchants. 'The skills that Piero or any painter used to analyse the forms he painted were the same as Piero or any commercial person used for surveying quantities,' Baxandall writes. 'The literate public had these same geometrical skills to look at pictures with: it was a medium in which they were equipped to make discriminations, and the painters knew this.' So when Piero placed a pavilion behind the Virgin in his 1460 work *Madonna del Parto*, he knew his audience would have a firm grasp of volume and shape.

Given the glamour surrounding the clipper races, it is not surprising that tea clippers have inspired dazzling canvases and prints. Nineteenth-century artists such as Samuel Walters and Robert Salmon worked around ports such as Liverpool and London and produced portraits of tea clippers in the way other artists painted portraits of people. However, it was twentieth-century British painter Montague Dawson who, looking back at the clipper era with a nostalgic eye, portrayed the vessels in a way that really captures the drama and romance of their voyages. His images of swirling sails and turbulent seas – which now fetch prices that run to hundreds of thousands of dollars – have firmly established the tea clippers' place in the maritime painting genre.

More recently, American pop artist Andy Warhol seized upon a less romantic image through which to express his view of twentieth-century American culture: the tin can. When Warhol's *Campbell's Soup Cans* were first shown in a Los Angeles gallery in 1962, the soup can paintings enraged a nearby gallery, which installed a display of actual soup cans in its showroom and charged buyers twenty-nine cents a can (Warhol's works were going for $100). When asked why he painted soup cans, the artist reportedly replied that he ate soup every day for lunch – or perhaps it was more along the lines of something he wrote in 1962: 'My image is a statement of the symbols of harsh, impersonal products and brash materialistic objects on which America is built today. It is a projection of everything that sustains us.'

The shipping container, too, has crept into the creative consciousness. These steel boxes that have travelled so many miles, often with food inside them, have caught the imagination of contemporary architects who have turned them into unorthodox but chic homes. In 2001, British architects Nicholas Lacey & Partners completed a scheme, appropriately enough in London's docklands, using recycled shipping containers painted in bright reds, oranges and yellows. Linked together they provide four floors of stylish but affordable apartments and work studios. In 2002, photographer Gregory Colbert came up with the ingenious idea of exhibiting his photographs around the world in a temporary space designed by architect Shigeru Ban and constructed entirely of shipping containers stacked on top of each other. The 'Nomadic Museum' now travels all over the world – and ships itself.

If our moveable feasts have sparked conflicts, fed global empires, altered the taste of foods and revolutionised eating patterns, they have also helped spread decorative ideas through the ancient world, advanced artists' theories of perspective and provided the inspiration for highly collectible paintings. Yet no transport innovation has had quite such a profound impact on art, architecture and design as the grain elevators, changing profoundly the characteristics of the buildings in which we live and work.

In 1946, New York's Museum of Modern Art published a booklet that was designed to introduce readers to modern architecture. It was a time for reflection on a style of architecture that by then was a well-established artistic movement. What is so interesting about the booklet – titled *What is Modern Architecture?* – is that several passages in the text could be applied as easily to grain elevators as to contemporary homes and offices. 'Much of the beauty of a modern building depends upon the broad, continuous sweep of flat exterior walls,' the authors wrote. 'This effect is often facilitated by the smooth perfection of machine-finished materials and enhanced through the deliberate and purposeful use of repetition.' One only has to think of the ranks of cylindrical towers and great expanses of wall that characterise Buffalo's industrial structures to make the connection between the elevators and the designs of modernist architects.

This monolithic, stripped-down style came to dominate twentieth-century architecture. In some cases, it was highly successful. In the USA, the International Style, the American form of Bauhaus architecture, produced some iconic buildings such as the 1949 United Nations headquarters in New York, whose design team included Le Corbusier and Wallace Harrison – and which is now considered by many as a modern masterpiece. Yet in 1960s Britain, the extreme utilitarianism of modernist design was deployed in the construction of cheap housing, creating soulless concrete towers that rapidly became dilapidated and riddled with crime. Le Corbusier is known as much for his theories as his buildings, and his work has come to symbolise the brutally functional outlook of the modernists. People often blame the renowned French designer for the grim concrete structures, bereft of ornamentation, that characterise much of 1960s architecture. But how many are aware that one of his inspirations was something designed for transporting food?

12

Going with Gravity

Cold War weaponry finds a new purpose

Maize: A cereal grain, also called corn, from the species *Zea mays*

Origin: Mesoamerica

Etymology: Spanish *maíz*, from Taíno *maisí, mahis*

Legends: The Aztecs believed that maize was on the earth before mankind. The god Quetzalcoatl changed himself into an ant to retrieve kernels of maize from the Mountain of Sustenance. He then took it before the Tamoanchan gods who agreed that corn should be passed on to humans

*I*N THE GRIMY interior of a cargo plane, a young Russian is making coffee. He has a pale, slightly sallow complexion, heightened by the harsh light inside the aircraft. His scrawny torso is covered by a faded black T-shirt and his jeans are slightly too large. He opens a large tin of Nescafé – the words 'Kick-start Your Day' are splashed across its label – scoops a few granules into a mug and pours hot water on top. After taking a sip, he lights a cigarette and stares blankly out of the window as the plane makes its way over the Kenyan border and into Sudan. After a few minutes, he stretches out for a nap on a couple of the battered passenger seats in the small cabin.

Stashed in the hold behind him is a precious cargo – ten tonnes of maize and sorghum packed in white sacks stamped with the United Nations World Food Programme logo. Each bag carries 110 pounds and has been encased in four outer plastic sacks, protection against the impact of the impending fall. Until then, heavy chains hold them in place on the floor. It is February 2006 and this is flight UN-A156. In what is the last leg of a voyage that started thousands of miles away on a farm somewhere in the middle of America, these sacks of food are heading for their ultimate destination: the scorched terrain of South Sudan.

The food is travelling in an Antonov 12, built almost fifty years ago, at a time when African nations were valuable tools in Cold War strategies. The AN-12s went into service in 1957 (the year the Russians shook the world by launching the first artificial satellite, *Sputnik*). Designed to shift Soviet weaponry, this four-engined turboprop is an impressive machine, known for its upswept rear fuselage, stepped cockpit and glassed-in nose. Inside this particular model, homely touches introduced over the years have softened its hard military edges – a plastic container of toothpicks secured to the wall with rubber bands, a set of mugs and an old wooden stool that rattles violently on takeoff. Flimsy nylon curtains decorated with hearts and flowers hang limply at the windows and the plastic panelling lining the walls is a sorry imitation of wood. In a bored moment, someone has outlined the image of a tiny aeroplane by sticking pins into a section of the green felt above one of the dust-covered windows.

What might the Soviet designers of this machine have thought had they seen their old warhorse thus transformed? In 1989, this machine and others like it became part of Operation Lifeline Sudan, an emergency humanitarian operation that was established to feed millions of starving people in the midst of a conflict. The war pitted the Muslim government in the north against the rebels representing the largely Christian and animist south (a separate brutal conflict in Darfur continues to rage). Although a fragile peace descended on the south after the signing of an agreement in early 2005, divisions and hatreds persist. Delivering food remained essential even after the signing of the peace accord – for while the conflict itself was brutal enough, war also destroyed the region's infra-structure. Local agricultural production was further obstructed by a series of devastating droughts. Many areas of southern Sudan remained hard to reach, with mines a constant hazard to ground vehicles and roads that during the rainy season turn into rivers of mud.

Airdrops were the solution, and the Antonovs were among a handful of aircraft left in the world able to do the job. With military requirements changing and modern planes built to new specifications, the old Antonovs – along with their American equivalents, the C-130s – remain among the few machines capable of delivering food aid. Yet memories of the plane's battlefield days are alarmingly fresh. At the height of the conflict, Sudanese Air Force Antonov 26s were accused of attacking civilian targets. Villagers soon became adept, say aid workers, at distinguishing between the engine sounds of UN food planes and the low drone of the approaching bombers. Some even say it was the same crews dropping bombs one day and food the next – an apocryphal tale, perhaps, but not beyond the realms of possibility. After all, both missions were being executed by Russian planes and, with operating instructions on board written in Cyrillic, only those who can read Russian can fly these aircraft. Moreover, since all crews needed Sudanese visas and any planes entering Sudan had to be registered with the government, only a limited number of Russian or Ukrainian aircrews were able to fly within the country.

Apocryphal or not, such tales epitomise the bizarre contradictions and dilemmas that often surround the task of delivering emergency food aid – and in Sudan, the issues are as muddy as those roads during the rainy season. When British and American pilots flew food into Berlin during the 1948 blockade, they knew what they were fighting for – freeing the city from Stalin's clutches. In feeding the starving Sudanese, the United Nations, charities and non-governmental organisations were not using food as a bloodless weapon in a clear-cut conflict. They remained politically neutral, mopping up while the bullets were still flying in an attempt to prevent a humanitarian disaster that was born of a messy internal war.

Even the arrangements necessary for the supplies to be delivered were wrought with difficulties, since in order to allow the UN food planes into Sudanese airspace, access had to be negotiated with both the Sudanese government and the rebel groups – a process that forced each side reluctantly to acknowledge the other as a political entity. In a ground-breaking agreement for the humanitarian industry, the warring factions met in Switzerland, along with United Nations representatives and other officials. During the process, so the story goes, former schoolmates from Khartoum and Juba university alumni who had not seen each other for years would go drinking at night in Geneva's bars and, the next morning, would be fighting at the negotiating table.

If the plane being used on flight UN-A156 comes from a fleet with dubious credentials, the crew are unlikely heroes, too – burly Russian contract pilots who seem indifferent to the business of relief work. In the cockpit they sit with large green headphones over their ears facing a control panel bristling with dials. One of the team wears a T-shirt with the puzzling message '1933-Nowadays. Philosophy of Conquest' printed over a map of Europe beside an eagle encased within a golden circle. After more coffee and cigarettes, he and a colleague embark on a game of poker using the large plastic cooler containing lunch as a card table. This is not among the images that spring to mind when picturing the world of humanitarian aid. Where is the wise old African doctor? The green young volunteers, burning with ambitions to change the world? Not here in this plane.

The place to find those people is on the ground. There, aid workers walk for hours on end from village to village to distribute supplies and assess the availability of local food sources. Anything that can be grown locally is a cause for celebration for severe under-funding is a constant worry for the World Food Programme, which at times is forced to reduce its rations. This is when tough decisions must be taken on where exactly cuts can be made. WFP rations for malnourished children and nursing mothers are rarely reduced. For others, however, the non-cereal part of the daily ration might be halved to stretch supplies, cutting precious calories from an already meagre diet. These are the sorts of decisions aid workers dread. They have seen the results of malnutrition. More than 800 million people in the world are chronically hungry. While these people are not starving, the effects of 'the hunger' are debilitating. Children's growth – both physical and cognitive – is hampered, mothers give birth to underweight babies, adults lack energy and, with a lowering of the body's resistance, disease is a constant danger.

The crew of flight UN-A156 will not be encountering this side of the aid industry. They require no visas or luggage for their trip. After all, the flight will not be landing. Once the aircraft has shed its load, it will turn round and head straight back to Kenya to pick up the next consignment of food. For the pilots of this flight, the only sight of Sudan is a distant one from the windows of the plane. It is a view that consists of nothing-ness. Leaving behind the airstrip in Kenya, the plane crosses the Sudd, vast swamplands that spread outwards from the banks of the White Nile as it winds south from Khartoum through a country the size of western Europe. Droughts have robbed the terrain of moisture. Mile after mile of brown earth is interrupted only by small huts within tiny enclosures – perfectly formed circles that are the sole evidence of human habitation. On closer inspection, it is possible to detect the odd figure; a group of children; a few cattle lodged beneath dry, shrubby trees.

Like the families of West Berlin in 1948, the villagers living in these settlements depend on aircraft for their food. However, their supplies are not unloaded by cargo handlers on the ground. What sustains them comes hurtling down towards them through the sky. While airlifts were used in the early days of Operation Lifeline Sudan, with supplies flown to any place with an airstrip, this encouraged villagers to leave their homes and set up camp around the airstrips, creating large communities of displaced people. Airdrops could distribute food nearer to people's homes. Moreover, since the planes never touched the ground, the fleet was subjected to far less wear and tear. For almost two decades, Operation Lifeline Sudan – which acted as an umbrella group for the relief efforts of aid agencies and NGOs – used this method to deliver staples such as maize, sorghum, salt and sugar to thousands of villages. It has proved one of the world's biggest humanitarian airdrops. At the height of the crisis in 1998, eight planes loaded with UN sacks were each making two or three sorties a day from Kenya into Sudan. An even larger fleet of light aircraft transported the teams working on the ground.

Their starting point is Lokichoggio. This eccentric little nowhere town sounds more like an unpleasant tropical disease than a centre for aid distribution. Affectionately known as Loki, it lies at the foot of the Mogila Mountains in Kenya's far north-west corner, about twenty miles from the Sudanese border. From the air, it is not much to look at. Mud huts, haystacks and low-rise buildings with tin roofs are strung out along the airstrip. Small bushes and trees spread across the arid terrain like stubble on an unshaven face. On the ground, dusty tracks pass for streets and a ramshackle collection of provision stores and bars go by names like Tuff Gong Merchants, Upland Fashions and Pub Half London.

This is the unprepossessing setting for what until recently has been one of the world's most important humanitarian logistics hubs. In 1989, as the crisis in South Sudan escalated, everyone from Médecins Sans Frontières and World Vision to UN agencies such as the World Food Programme, Unicef and the UNDP rushed here, installing themselves in compounds behind tall wire fences. Private entrepreneurs set up camps and hostels providing accommodation for the new arrivals. The population soared. A forgotten outpost with nothing but an airstrip, a Christian mission and a shop run by a Somali trader became a crowded logistical nerve centre with a population of nearly a thousand – a small town almost entirely made up of aid workers.

Outside the wire fences surrounding the humanitarian compounds, small children on large bicycles battle through the hot wind outside the Loki Light Academy. 'Rise and Shine' is the school's motto, painted on a wooden sign below a picture of a light bulb – or is it the sun? Members of the local Turkana tribe are among the pedestrians on this desolate road. The men, their arms draped decorously over sticks across their shoulders, wear hats that are slightly too small and vaguely resemble trilbies (no one seems to know where they acquired these). Women in earth-coloured shawls wear thick beaded neck decorations. Tall, thin and beautiful, they seem to have a haughty elegance and an otherworldly air – until they approach you saying: 'Give me fifty Kenyan shillings, please.'

In Loki, everything revolves around transport. Roaming about the place are large white vehicles, the essential accessories of the humanitarian world. The shipping container pops up everywhere, with relief organisations occupying the steel boxes that once carried cargo. Over at the UN compound, an eccentric thatched version is the World Food Programme's radio room and the headquarters of Belgium's Veterinarians Without Borders combines a shipping container on the ground floor with an upper storey resembling a Swiss chalet. When they are not inside these makeshift offices, relief workers can be found at the airport. Loki's forty-year-old airstrip is the only piece of infrastructure that pre-dates the aid operations.

It is a busy little airport. Passengers sit on cheerful blue benches with the letters WFP painted on them to wait for their planes to depart. Kenyans with plastic UN badges on chains round their necks check messages on their mobile phones. South African pilots smoke a last cigarette. On one side of the runway, large trucks arrive from Mombasa with sacks of grain, which are unloaded and stacked inside 'rubbhalls' (short for rubber halls, the large tents that constitute the aid industry's mobile warehousing). Nearby are the United Nations aircraft, painted in white, with the letters 'UN' as their simple livery. The strip of tarmac 20 metres wide receives a constant stream of traffic – everything from small passenger planes bringing relief workers up from Nairobi to the heavy lifters that lumber over the border to take food into Sudan.

As this particular heavy lifter, flight UN-A156, heads northwards, villagers on the ground in South Sudan are preparing the daily meal – a pot of *ugali*, a stiff porridge made with maize meal. It is a meal behind which lie careful calculations. Before an airdrop takes place, the World Food Programme conducts a 'needs assessment' to ascertain how much food is available locally and how much it needs to deliver. Villages are notified in advance of the date and location of the airdrop and, on the appointed day, family members set off on treks of up to 6 miles to the drop zone. They carry with them the ration cards issued when they registered to receive food supplies. Once on the ground, the sacks of food are neatly stacked together and a team of local chiefs and relief workers carry out 'proportional piling' (deciding how much should be allocated to each village). Then the piles are divided once again into family and individual rations.

Once collected and taken home, the new supply of food is stored in the back of every small hut. Women take out a few handfuls and begin the long process – up to four hours – of grinding it into flour. Soon fires are lit and some water is brought to boil in a pot. The flour is added to the water gradually, requiring constant stirring while the mixture starts to thicken. Eventually, the maize flour congeals into a rubbery mass. The recipe could hardly be simpler. And yet before the family of South Sudan can finally sit down to this humble meal, its ingredients have been on a long, complex and often controversial journey.

The main ingredient, the maize, started life on a vast farm in the heartland of America. There in Ohio or Illinois, deep nitrogen-rich soils, hot sun and even rainfall provide ideal conditions for raising grain. From these fields, giant companies such as Cargill, Archer Daniels Midland and Louis Dreyfus – some of the world's largest agribusinesses – sell a percentage of the grain they harvest to the United States Agency for International Development (in 2004 this grain accounted for about half the $700 million USAID spent on its food aid programme). The maize is taken by rail or on trucks to American ports such as Lake Charles, a large facility in Louisiana where it makes a short stopover in a pre-positioning warehouse or is loaded directly on to the vessel.

Before the ship has even left the docks at Lake Charles, however, it has been generating controversy. While the USA is by far the largest food aid donor – in 2005, it funded almost half the food aid deliveries in the world and its donations were double those of the European Union – some argue that its approach is inefficient and expensive. Christopher Barrett and Daniel Maxwell, two food aid experts, have estimated that it costs the USA more than $2 in taxpayer money to deliver every dollar's worth of food aid.

When the US food aid programmes were developed in the 1950s, feeding the hungry was not the only motivation. As the Berlin Airlift had demonstrated, food could deliver political objectives. This theory was pursued as the Cold War proceeded. Through donations of food, the USA hoped not only to alleviate poverty and foster development but also to win the hearts and minds of people in developing countries – who might otherwise be tempted to form alliances with the Soviet Union. Hubert Humphrey, a US senator, summed up many of the motives of the time in 1953, when he said: 'Wise statesmanship and real leadership can convert these [food] surpluses into a great asset for checking communist aggression. Communism has no greater ally than hunger; democracy and freedom no greater ally than an abundance of food.'

In the 1950s, the food aid programmes had neatly solved another problem – agricultural surpluses that were piling up as mechanisation raised production and demand for food aid from Europe fell as the continent recovered from war. In addition, planners reckoned that if, through food donations, overseas markets could acquire a taste for American wheat and meat products, those markets might eventually become buyers of US food exports (little evidence has emerged to support this theory, according to Barrett).

Today, while its aims have shifted, the USA is still clear about the motives behind its food aid programmes. As the authors of *Foreign Aid in the National Interest*, a 2002 USAID report, write:

> Pre-empting threats and disasters is not the only reason that fostering development is in the US interest. Successful development abroad generates diffuse benefits. It opens new, more dynamic markets for US goods and services. It generates more secure, promising environments for US investment. It creates zones of order and peace where Americans can travel, study, exchange, and do business safely. And it produces allies – countries that share US commitments to economic openness, political freedom, and the rule of law.

Some have accused the USA of using its food aid donations for less laudable purposes – to prop up its heavily subsidised agricultural industry. However, this supposition turns out to be a myth. While some agribusinesses certainly benefit from the extra sales, the value of humanitarian food purchases made by the US government represents a tiny proportion of overall farm produce – about $2 billion a year in a country where an annual $900 billion or so is spent on food.

In fact, the docks are where one of the greatest inefficiencies of the US food aid system is found – in the vessels that will carry the maize to Africa. An obscure law – the 1954 Cargo Preference Act – requires that seventy-five per cent of all agricultural commodities sold or donated by the US government travel on American ships (which receive generous government subsidies), even if using foreign vessels is cheaper. The trouble is, the US maritime industry is in decline. It only handles about three per cent of all goods arriving at and leaving American shores (the rest is transported by foreign vessels). Thus there is little competition between the few maritime companies that qualify to bid for US food aid shipments. This pushes prices to levels that are substantially higher than the cost of sending the food on the same routes in foreign vessels and prompts frequent calls for reform of the system.

Leaving such debates aside, our maize sets out on its long journey from the USA to Africa. The sacks are part of a shipment of emergency food aid – as distinct from 'programme food aid' (the transfer of food from one government to another as a form of economic support) or 'project food aid' (food provided for development or nutrition programmes). From Lake Charles, the vessel passes through the Gulf of Mexico, through the Caribbean, across the Atlantic, round the Cape and up to the east coast of Kenya to Mombasa, a trading town since the twelfth century whose impressive fort is a reminder of the presence of the Portuguese in the sixteenth century. About twelve vessels a month arrive at Mombasa with food destined for countries in the region. The documentation needed to discharge and ensure clearance of food aid has been prepared weeks in advance to ensure there are no delays that will slow its delivery. As soon as the ship has docked in Mombasa harbour, grain starts pouring out of the side of the vessel. The maize is put into paper sacks at a bagging facility near the port. Then it is packed into sturdy trucks which, travelling in

going with gravity

convoys for reasons of safety, set out on the week-long voyage to northern Kenya and Lokichoggio.

Once on this side of the world, the grain is far from free of difficulty. Local leaders and national governments often control or manipulate charities and aid agencies, while warring factions and local strongmen hijack food aid and use it to further their military aims, either simply stealing the food or pretending to be civilians. Human Rights Watch has documented the way that, during the war in southern Sudan, government army officers profiteered from relief food while rebel forces imposed 'taxes' (often consisting of part of their relief food supplies) on civilians and diverted food aid for their soldiers before it reached the people it was supposed to feed. At one point during the war, some said that food aid was actually prolonging the conflict.

If food aid does not fall into the hands of rebels, other debates surround its delivery. A bitter row erupted in 2002 when, despite the threat of famine, the Zambian government refused to accept US food aid that had been genetically modified. It claimed America was using aid to introduce genetic modification to Africa by stealth. Andrew Natsios, USAID director, accused environmental groups of putting people's lives in danger by denying them vital food supplies. 'They can play these games with Europeans, who have full stomachs, but it is revolting and despicable to see them do so when the lives of Africans are at stake,' he told the *Washington Times* in August 2002.

Regardless of whether or not genetically modified food poses health risks, other risks are attached to the acceptance of GM food aid. Countries hoping to sell their farm produce to European countries must guarantee that it is GM-free.

Moreover, European Union financial assistance for developing agriculture would be denied to nations growing GM crops. This would mean erecting expensive safeguards to prevent the spread of the seeds from GM food aid to local crops. While other African countries had initially resisted the GM aid, Mozambique, Lesotho, Malawi and Zimbabwe agreed to receive it if it was milled before reaching their borders. But Zambia refused. Adding to the 2002 controversy were pro-GM and anti-GM groups in the USA and Europe – yet another demonstration of the fact that food aid is often caught up in broader political wrangles that take place far from the homes of the hungry.

Controversy also surrounds the impact of food aid on local markets. Some argue that shipping in imported food during a crisis can push down the price of agricultural produce locally, leaving farmers with no income to buy the following season's seeds. On the other hand, buying food locally in an area where scarcity prevails could drive up prices to such an extent that no one can afford to buy it. The United Nations now tries to buy food from nearby countries wherever possible, since this can even help stimulate economic activity if managed properly. The creation of food dependency is another criticism levelled at the aid industry. Aid workers point out that nomads in South Sudan who would traditionally have moved to greener areas during the dry season now remain in one place year round. In short, the nomads know that they will be fed by the food that falls from the sky.

Then there is the cost of delivery. Airdrops are an extremely expensive way of delivering relief supplies. The 'block hour rates' (covering not only flying time but also things such as loading time, fuel and airport fees) for the various aircraft contracted to Operation Lifeline Sudan are high. For the C-130s, it is a hefty $4,862. A Buffalo (an aircraft used for airlifting supplies such as vegetable oil, for which landing is required) costs $2,464 an hour. At $3,492 per hour for the Antonov, its three-hour flight will notch up more than $10,000 – and the plane might make three of those journeys a day. While the World Food Programme negotiates competitive deals with its contractors, during the Sudan crisis aviation accounted for a huge budgetary chunk of a relief operation that some said at its peak cost the international community about a million dollars a day.

Of course, for the World Food Programme, the cost of airdrops means they are a last resort, used only when conflict prevents delivery or where infrastructure on the ground has broken down. 'I don't like using the air,' says Amer Daoudi, who has acted as the WFP's co-ordinator in many emergencies. 'When all other options fail you resort to air – and I'm stubborn. I won't let all other options fail, so I always try to deliver by conventional means.' Daoudi's use of the word 'conventional' is an intriguing one. In fact the WFP is intensely creative in devising the means to get its food to the hungry. 'We've used donkeys in Sudan, we've used elephants in Cambodia,' Daoudi explains. 'We have barges in Sudan – in 1991 we did the first ever barge convoy between Malacale in Sudan and Tuva.'

Daoudi is an erudite Jordanian who has worked for the WFP for almost fourteen years. While he talks, long, elegant fingers move through the air as if he were tracing the movements of a classical ballet. At times, he sweeps a hand across his desk, as if planning an imaginary military campaign across some invisible map. Sitting in his office in an immaculately pressed suit, he looks like a diplomat, a politician or a senior business executive – someone who might have spent his life in conference rooms and international hotels. This could not be farther from the case. Daoudi is a hands-on logistician who, like so many of his colleagues, has worked in hair-raising circumstances in countries where war, famine or natural disasters have created chaos and starvation. His passion emerges as he talks. 'We are faced with landmines, wars, banditry, rebels. Sometimes the local governments do not even exist. And ninety per cent of our operations take place in areas without any infrastructure, so it's a process of continuous improvisation,' he explains. 'We work in circumstances that no commercial company would dare venture into. We can't say: "Oh, there's a war and there's bombing – we can't go." No, people need food. So we've got to find a way of getting it to them.' He stretches his arms wide as if welcoming the world's hungry into the safety of his personal fortress.

The comparison with the commercial food delivery sector is an interesting one. As the corporate world looks for ways to help alleviate global disasters, transport and courier companies are forming philanthropic partnerships with relief agencies, donating time and expertise rather than money. FedEx has long worked with international relief organisations such as the American Red Cross, donating its transport services to help with the distribution of food and other emergency supplies. DHL is working with United Nations organisations to help cut bottlenecks at airports near the scene of disasters as well as assisting the International Federation of Red Cross and Red Crescent Societies in developing disaster management tools. TNT Logistics, part of a Dutch logistics group, is helping the World Food Programme with things such as computer cargo tracking systems and management of its fleet of trucks. These partnerships make a lot of sense. After all, both aid agencies and commercial logistics companies are constantly striving to achieve greater speed and efficiency.

Yet the logistical obstacles to delivering food in times of war or natural disaster make getting those Thanksgiving turkeys or barbecued ribs to customers look easy. As Daoudi points out, local infrastructure has often been destroyed, leaving nowhere for cargo planes to land and no roads for food trucks to follow. In conflict zones, rebel forces may try to obstruct the passage of supplies. And while food companies can anticipate a rush on turkeys and hams at Christmas or chocolate eggs at Easter, relief agencies have to deal with sudden surges in demand for food whose timing, location and scale are unpredictable.

Getting hold of the food is a huge challenge, too. Many aid organisations have grain and other staples stockpiled in warehouses round the world, but since the site of the next event is usually impossible to determine, those supplies may still have to be transported great distances, at great speed. With consignments of food aid constantly travelling across the world's oceans, vessels must often be rerouted towards the latest humanitarian disaster. While companies have warehouses in places near the markets where they sell their goods, aid agencies must set up distribution centres virtually overnight, usually in extremely difficult conditions. Moreover, while for companies, late deliveries may generate a few angry phone calls from customers, for organisations like the World Food Programme, delivering on time means the difference between life and death.

The WFP's mandate is to deliver food to hungry people, whatever the circumstances, through whatever means possible

– whether that is on the back of a camel or in the hold of a Russian aircraft. In the process, the organisation has prevented an astonishingly large number of people from dying. In 2005, it delivered food to ninety-seven million people, thirty-five million of them caught up in wars, famines, epidemics or natural disasters. During the 1992 drought in southern Africa, the WFP saved eighteen million people from starvation – equivalent to the populations of Switzerland and Belgium combined. The sheer tonnage of food moved by the WFP is mindboggling. Between 1962 and 2001, it shifted almost 70 million tonnes of food to a hundred countries – the same weight as almost 192 Empire State Buildings.

For those in charge of delivering food aid, answering the questions thrown at their industry is not easy. Creating a culture of dependency is a problem acknowledged by many in the aid business. Yet in 1948, the Universal Declaration of Human Rights included the right to food among its thirty articles. With a mission to fight starvation – part of the United Nations' commitment to fulfilling this human right – the World Food Programme and other relief agencies cannot cut off the people they were created to serve. In the business of hunger, successful exit strategies are hard to find.

Round the dinner table in Lokichoggio, the debates are hard to escape as relief workers offload the frustrations of the day. 'I don't understand it – these people are stealing from each other!' exclaims a young Australian nurse one evening, describing her dismay at seeing World Food Programme relief supplies being sold in a local shop. 'Why are we here, if that's what they're going to do?' Sitting opposite her, a Sudanese member of the Carter Foundation team looks on with sombre detachment. He has no doubt heard it all before. 'You have to understand,' he finally begins in a quiet voice. 'This war went on for twenty-two years. For some people, it's the only thing they've known. I remember when that sort of behaviour would have been shocking to the Sudanese – but during a war, people do whatever it takes to survive.'

At this point, pudding is served – a particularly fine crème caramel. Good food and leafy surroundings have made Trackmark camp the preferred accommodation in Loki. Here, the day ends with drinks at the bar, followed by a buffet dinner and perhaps a few moments relaxing on large cushions in front of the widescreen television set to catch the news from CNN or BBC World. On Saturday nights, barbecues are held by the small swimming pool. 'When I first bought this land, it looked like that,' says Heather Stewart, pointing to the desert that lies outside the camp. An enterprising bush pilot who has been in Loki for more than twenty years, Stewart turned her barren piece of land into a small oasis, replanting trees from surrounding areas to bring shade to her well-watered gardens, now bursting with bougainvillea and desert roses. With a beach-style bar, a small conference centre and a cybercafé, the place would pass for a simple but pleasant resort – except that, instead of tanned tourists, guests include hard-bitten de-miners, medics and logisticians. Every morning after breakfast, they pace up and down on the lawn, glued to sat-phones, trying to track down their drivers or ascertain the whereabouts of the latest delivery of supplies.

Camps such as Trackmark have also attracted criticism, the press accusing relief workers of living in luxury while Sudanese starve next door. Others disagree. 'It's ridiculous,' says Bill, a Scottish engineer living in Loki. 'Where do they expect these aid workers to live?' He has a point – it is rather like demanding that a doctor treating a sick patient should be forced to catch the disease himself. But wherever the humanitarian industry has a presence, these tensions are never far away.

The humanitarian world is on the move again. The advent of peace in the south means agencies can operate independently within Sudan and many are moving their offices to the banks of the Nile at Juba, the southern capital. Loki is growing quieter. After almost twenty years of dropping food across the border, the World Food Programme is winding down its air operations here. As well as providing food assistance for the South Sudanese as they make their slow recovery from war, the UN is busy building roads in Sudan and removing mines from existing ones so that trucks can replace planes. Expensive food deliveries by air are being cut. Eventually, it is hoped, they will be stopped altogether. It is the tail end of a monumental operation that has seen thousands of tonnes of food dropped from UN cargo planes, feeding millions of people.

After ninety minutes in the air, flight UN-A156 is about to shed its load. The plane descends to about 700 feet from the ground, low enough to minimise breakage of the bags but not so low as to create horizontal thrust, which could send them bouncing across the ground, splitting the sacks and spilling valuable grains. Even so, villagers are waiting below with gourds, ready to collect the contents of any breakages (although the sacks of food are packed inside four tough nylon bags, they still occasionally break). Ants pick the harder-to-reach grains from the ground around broken bags after the villagers have finished. Relief workers have observed that during the 'hunger

going with gravity

season', when starvation levels are severe, people can be found following lines of ants back to their nests to retrieve the food that the creatures have stored there.

From the navigator's post in the Antonov's glassed-in nose, the drop zone below can be clearly seen – like an oversized football pitch waiting for players to arrive. Corners have been marked out with empty white UN sacks; a large cross is at its centre. Teams on the ground have been flown in to secure the area and prevent locals and their animals from straying too near the falling sacks. As the pilot negotiates his way towards the drop zone, two of the crew set about removing the chains securing the plane's payload. The back doors of the aircraft open. Sunlight floods the interior. One of the team counts down on his fingers. Four, three, two, one – and suddenly the back end of the plane lurches downwards. Coffee mugs slide across the table. With the aircraft's nose pointing towards the sun, gravity does its work, sending sacks sliding swiftly across the floor. In an instant, 10 tonnes of food are hurtling out into the air.

For a few seconds, these hefty objects seem possessed of a tremendous lightness. Like large white snowflakes, they flutter down from the sky until they hit land with an almighty thud. But in a brief and thrilling moment between the plane and the earth, the sacks of food appear to have been liberated from the controversies that swirl around them on the ground. Questions of food aid budgeting, arcane shipping laws and genetic modification suddenly seem far away. For the families that will soon be arriving to collect them, these well-travelled packages have become something simple, profound and essential – the main ingredient in tonight's dinner.

epilogue

ONIGHT'S DINNER — IT sounds so simple. Yet whether 'dinner' is a bowl of maize meal *ugali* or a pan-fried Norwegian salmon fillet, accompanied by sautéed French beans and washed down with a bottle of Californian chardonnay, its ingredients have been shunted around by a vast array of machines and vehicles before reaching the table. Along the way, people, technologies and systems have participated in a global battle against time, geography and the seasons to bring us strawberries on a chilly December morning or mangoes whose life started on a tree in southern Thailand. It is a mammoth global relay race. Foods and drinks are passed swiftly along each section of long and complex chains before reaching their destinations. Eating is the easy bit. Each mouthful is the last step in the journey (unless you consider what happens to food after it has passed through the human body — but that is another story).

Some lament the fact that the things we eat accrue so many 'food miles' and, indeed, journeys like those of the fish sent to China for filleting before returning to Europe or the USA seem bizarre – if not downright insane. And while processed foods may not have been to China and back, their various components may have visited an alarming number of different destinations as they are sorted, processed, packaged and distributed, creating dense transport networks like those mapped out by Stefanie Böge in her studies of yoghurt. The diversity of fruits and vegetables is diminishing as the industry focuses on robust, long-lasting varieties rather than those full of flavour and nutrients. Tomatoes, which bruise easily, are picked green and firm, shipped to their destination, and then sprayed with ethylene gas, which artificially ripens and reddens them but leaves them crunchy and tasteless.

Clearly, transporting food does not always improve the food itself. Yet without the ability to move our feasts, it would be a lot harder to fill the shopping basket described at the start of this book. Take the olive oil. For those in regions with hot, dry climates, the supply could be plentiful. Elsewhere, without a Roman amphora or a stainless steel tanker, cooks would have to return to using lard, and salad dressing would be a thing of the past. Norwegians would be able to get fresh salmon, as would the Scots, Alaskans and others living near rivers. Fish farming would help. However, without an ocean-borne shipping container, filleted salmon might cost a lot more. The British may be experimenting with home-grown tea, but its cultivation requires a mild climate with no severe frosts. Without canning, the San Marzano plum tomatoes would have to be eaten in San Marzano. Our tub of low-fat yoghurt would rely on the presence of a nearby farm, as would the flour.

Strawberries would be available only at certain times of the year and the bottle of chardonnay from Napa Valley would grace the tables of Californians alone. Few of us would have set eyes on a banana, have tasted curry or have chewed gum. The corn on the cob, one of the great staples of the global food supply, would be the most widely available food on the list. It seems unlikely that many of us would be prepared to relinquish our global cornucopia in favour of a diet based entirely on what grows near home.

However, as food continues to circle the globe, some worry about the impact its journeys are having on the planet. With the threat of global warming becoming more apparent, food miles have been targeted, as a database search of newspapers and trade magazines reveals. Between 1999 and 2007, the term 'food miles' appears in nearly 3,000 articles (many of them in British publications), with talk of dinners 'washed down with tanker-loads of diesel' and meals that 'cost the Earth'. In short, globetrotting foods stand squarely accused of destroying the planet.

The concept of 'food miles' has certainly helped raise awareness of the environmental impact of one aspect of our daily lives: eating. However, is this really the right fight to pick? Calculating carbon emissions is a complex business, often confounding expectations. The buildings we live and work in (as well as those in which food is produced) generate, by some estimates, forty per cent of the carbon dioxide and other emissions that contribute to global warming. Food production, processing, cooking and car journeys to the supermarket all generate greenhouse gases. Eating local, seasonal produce is clearly a good option when it is available nearby. Yet the environmental benefits of doing so could soon be wiped out by

epilogue

regular trips in a gas-guzzling vehicle down to the farm shop. Shipping certain foods by sea from far-off countries such as New Zealand, where livestock can graze outside year round and sunshine is plentiful, might actually do less harm to the planet than obtaining them nearer to home if fertilisers are used in larger quantities, vegetables are raised in hothouses and animals are housed for part of the year, and need fodder and bedding (which require energy to harvest and produce).

In any case, equating the distance travelled by food with damage to the environment does not provide an accurate picture of the 'carbon footprint' of food transport. This varies wildly according to what types of vehicles are being used and how much they are carrying. So while fruit may have travelled halfway round the world in a vessel powered by fossil fuel, if that vessel is the *Emma Maersk*, she is carrying an enormous load (remember, a 20-foot container holds about 48,000 bananas and the *Emma Maersk* carries up to 11,000 containers) and she is powered by an engine which, by recycling the exhaust, increases efficiency by up to ten per cent and reduces emissions and fuel consumption. A more realistic assessment might be how many pounds of carbon dioxide are generated by transport for each pound of food consumed. By this measure, the *Emma Maersk* does relatively well while a 10-mile trip by car to the farm shop to pick up a bag of locally grown potatoes and a few carrots starts to look, in environmental terms at least, downright destructive.

The advance of civilisation has depended on being able to convey food from where it is grown or produced to shops, kitchens and dining rooms. The Incas knew this. While, by the fifteenth century, their empire remained primitive in many ways, they had one bit of it extremely well organised: the food supply. To manage territories equivalent to those stretching from Spain to Moscow, they built tens of thousands of miles of roads (carving into the Andes where necessary) and stored surplus grain in a series of imperial warehouses. Supplies from one part of the country were used to relieve shortages in another. Travelling in relays, teams of *chasqui* runners – athletic messengers wearing feathered sunhats and slings and carrying shell trumpets – sprinted along the roads to recite messages or deliver pictograms (the Incas never developed writing), covering up to 250 miles a day. This combination of an efficient communications system and an extensive transport infrastructure allowed the Incas to keep their people fed. It also enriched the imperial table. Using the *chasqui* runners, the Incas sitting in their palace at Cuzco, the empire's inland capital, could eat fresh fish caught off the coast at Chala, 200 miles away.

Compare this to Mozambique, where roads are so bad that in Maputo it is cheaper to import produce from Europe – where subsidies artificially lower its price, thus preventing the country from developing its own food exports – than what is grown by local farmers. Poor transport infrastructure is said to be among the biggest obstacles to progress in many African countries. Efficient transport, on the other hand, generates income and jobs, enabling countries such as Kenya to sell agricultural produce in western markets. Ensuring working conditions are acceptable and farming practices are sustainable is a challenge, but the existence of an industry that has created paid work for so many is surely an improvement on the alternative – aid handouts. For the South Sudanese, the chances of breaking free from the constant drip of food aid look remote for now. However, parts of their country are extremely fertile and, given roads and airports, the Sudanese, too, might free themselves from poverty – and from dependence on the UN planes that drop American grain from the sky.

Thousands of years ago, it was mankind that did the moving, chasing wild animals and seeking out the right berries to eat. It was called hunter-gathering. Around 9000 BC, the practice of agriculture emerged in what is known as the Fertile Crescent (covering parts of present-day Jordan, Iraq, Syria and Turkey). Domestication of plants and animals meant we no longer had to wander in search of food. Historians point to that moment as a landmark in the advance of civilisation. Where agriculture went, food transportation followed. Among the earliest uses of the wheel (often attributed to the Sumerians, who occupied today's south-east Iraq in 3000 BC) was on ox-drawn carts carrying farm produce. Farming surpluses plus the means to move them around meant people could engage in non-agricultural activities and live in settlements away from farmland.

During the Industrial Revolution, the dramatic fall in death and disease came not only as a result of advances in science and medicine. New transport systems and preservation techniques delivered larger quantities of food and more varied diets. As populations shifted from rural areas to towns, a moveable food supply became of critical importance. Joseph Dart, the nineteenth-century Buffalo entrepreneur, had grasped this. 'The road from hand to mouth is short and easy enough with men at first,' he wrote in his paper read before the Buffalo Historical Society in 1865, 'but as society grows, and division of labor is made, producers and consumers of food become widely separated, and the question of transportation becomes exceedingly important.'

Of course, Dart and his predecessors did not face the prospect of global warming. The Romans generated no carbon emissions with their ships, pots and slaves. The tea clippers relied on the power of the wind. Speed and efficiency was their concern – not how many tonnes of carbon dioxide were being released into the atmosphere.

Yet while we can certainly increase our consumption of local food, it seems unlikely that we can do without well-travelled produce altogether. For a start, it is not always practical. A BBC television series tracking Ollie Rowe's efforts to obtain all the ingredients for his restaurant, Konstam at the Prince Albert, from places within access of the London Underground highlighted some of the difficulties. In episode four, Rowe discovers that most of the traders at London's biggest fish market, Billingsgate, do not source their fish from the Thames Estuary and many buy it from outside the UK. The chef has trouble identifying suppliers that can deliver produce in sufficiently large quantities, and in episode nine (only three weeks before the opening of his new restaurant) he still has not found any locally produced spices or oil. Leaving practical difficulties aside, budgetary questions remain. Farm shop fare is often more expensive than what is available at the supermarket, so not everyone can choose locally grown fruits and vegetables.

Choices can be made, however, as to how we move food around – choices that could reduce the impact of long-distance haulage on the environment. Ocean-borne vessels are among the most carbon-efficient forms of freight transportation, for example. The *Emma Maersk* can move almost 50 miles using the same amount of energy per tonne of cargo that a jumbo jet uses travelling less than a third of a mile. Trains should, where possible, replace trucks. Companies could manage their logistics better, so that they are sending out full cargos rather than partially loaded trucks, as is often the case. The environmental performance of road-based vehicles can be improved. UPS, the package delivery company, uses routing technology to cut fuel consumption by devising journeys for its trucks that, for instance, avoid left-hand turns (which for US drivers mean idling in the middle of the road facing oncoming traffic). Associated Food Stores, a Utah-based grocery company that delivers to 600 stores throughout the Midwest, used the technology to cut back its fleet by thirty-eight per cent and reduce the annual journeys of its vehicles by more than 400,000 miles.

As cities expand, there are opportunities to capitalise on wasted space. Many believe the tops of apartment buildings and office blocks could contribute to urban supplies of fruits and vegetables. In New York City, the Earth Pledge Foundation has created a kitchen garden on its roof that in summer erupts with ripe organic tomatoes, peppers, cucumbers, sweet potatoes and herbs such as rosemary, basil and sage. In Vancouver, the Fairmont Waterfront Hotel has a 2,100-square-foot rooftop garden where it grows herbs and vegetables for use in its restaurant, saving the company about C$25,000 a year. Rooftop vegetable patches will never feed the world. But for city dwellers who crave locally produced fruits and vegetables, harvesting from the skyline would be a convenient option. For people who rely on cars, buying food from large, centralised sources such as street markets, farmers' markets or supermarkets burns less fossil fuel than driving around between different shops.

Better still, we need to find cleaner fuels for our vehicles. Some look to hydrogen fuel cells, which produce electricity from hydrogen and oxygen and whose only waste product is water. While ethanol has its doubters, others believe it holds the promise of a renewable and less polluting source of energy. In Buffalo, one enterprising businessman reckons that, as corn-based ethanol becomes more widely used, the old grain elevators could be woken from their slumber. Rick Smith has snapped up a set of disused elevators and plans to turn them into a large plant that will move and store corn for ethanol production, bringing lake freighters back to the Buffalo River. In doing so, Smith may achieve something generally thought to be impossible – to reclaim an infrastructure of the old world and set it to work in the new.

The new world is an increasingly virtual one. We can go grocery shopping on the internet. We can exchange recipes via e-mail, blogs or online social networks. We can watch web-based amateur videos of teenagers eating fast food. We can have our avatars, the digital recreations of ourselves, sit down to digitised dinner parties in the digitally designed homes of fellow participants in online virtual worlds. Yet we cannot eat what we find on the table in cyberspace. On the web, we can drink in ideas, but not liquids. In a world still rooted in materiality, we remain dependent on gravity-governed, old-world technology for sustenance. This is unlikely to change. Until someone develops a *Star Trek*-style transporter to beam solid objects across time and space, we will continue to rely on boxes, barrels, cartons, conveyor belts, ships, planes, trains and trucks – the linchpins in the peregrinations of our moveable feasts.

acknowledgements

As I researched and wrote this book, many, many people inspired, informed, entertained, advised, supported and encouraged me. Friends, strangers, colleagues and experts have lent me their ears, told me their stories, given me generous amounts of their time and graciously tolerated my curious obsession with cargo transportation. My heartfelt thanks to them all, and especially to Natasha Martin, my editor, and Zoë Waldie, my agent, as well as to Andy Barrons, Rachel Bennett, Tina Bennett, Paresh Chaudhry, Simon Crittle, Kenn Crossley, Gerry Darsch, my colleagues at the *Financial Times*, Gay Firth, Jeremy Gordon Hall, Ian Gowrie, Charis Gresser, Belinda Haas, Caroline Hurford, Rose Jacobs, Jonathan Jones, Beena Kamlani, Ash Khandekar, David Kuhn, Richard Lapper, Denise Lauer, Alessandro Lovatelli, David Miller, Gerry Mundy, Cait Murphy, Kate Murray, Tara O'Connor, Andy Pinning, José Remesal Rodríguez, Lynda Schneekloth, Jennifer Senior, Charlie Spicer, Frances Stonor Saunders, Nina Teicholz and David Wallis.

bibliography

Aaseng, Nathan, *Business Builders in Fast Food*, The Oliver Press, 2001

Abbot, Willis John, *Panama and the Canal in Picture and Prose*, Syndicate Publishing Company, 1913, Kessinger Publishing, 2004

Adams, Frederick Upham, *Conquest of the Tropics: The Story of the Creative Enterprises Conducted by the United Fruit Company*, Doubleday, Page & Company, 1914

The American Society Of Mechanical Engineers, *Thermo King Model C Transport Refrigeration Unit: An International Historic Mechanical Engineering Landmark*, 1 October 1996

Anders, Dr. Steven E., *The Best Feeding the Best: A Brief History of Army Food Service*, Quartermaster Professional Bulletin, Summer 2002

Antony, Joseph and Somasundaram, Rohith, *Benchmarking Supply Chain Framework: The Lunch Box Distribution Service in Mumbai*, National Institute of Industrial Engineering, Mumbai

Balakrishnan, Natarajan and Teo, Chung-Piaw, *Mumbai Tiffin (dabba) Express*, National University of Singapore, 30 April 2004

Banham, Reyner, *A Concrete Atlantis: US Industrial Building and European Modern Architecture*, MIT Press, 1989

Barrett, Christopher and Maxwell, Daniel, 'PL480 Food Aid: We Can Do Better', *Choices* magazine, 3rd Quarter 2004; *Food Aid After Fifty Years: Recasting Its Role*, Routledge, 2005

Baxandall, Michael, *Painting & Experience in Fifteenth-Century Italy*, Oxford University Press, 1972

Baxter, Henry H., *Grain Elevators*, Volume 26, Adventures in Western New York History series, Buffalo and Erie County Historical Society

Becker, Jasper, *Lost Country: Mongolia Revealed*, Hodder & Stoughton, 1993

Bell, J. Bowyer, *Besieged: Seven Cities under Siege*, Chilton Books, Philadelphia, 1966

Bernays, Edward L., *Propaganda*, 1928, Ig Publishing, 2004

Bernstein, Peter L., *Wedding of the Waters: The Erie Canal and the Making of a Great Nation*, Norton, 2005

Bielitzki, Dr Joseph, *Enhancing Human Performance in Combat*, US Defense Sciences Office, Defense Advanced Research Projects Agency (Darpa), 2002

Blasier, Cole, *The Hovering Giant: US Responses to Revolutionary Change in Latin America*, 1910–1985, University of Pittsburgh Press, 1985

Blázquez, J.M., *The Latest Work on the Export of Baetican Olive Oil to Rome and the Army*, Centro para el Estudio de la Interdependencia Provincial en la Antigüedad Clásica, 1992

Boehmer, Elleke (editor), *Empire Writing: An Anthology of Colonial Literature, 1870–1918*, Oxford University Press, 1998

Böge, Stefanie, *Road Transport of Goods and the Effects on the Spatial Environment*, 1993, Wuppertal Institute for Climate, Environment and

Energy; 'The Well-Travelled Yogurt Pot: Lessons for New Freight Transport Policies and Regional Production', *World Transport Policy and Practice*, Volume 1, Number 1, 1995

Booth, Martin (editor), *Opium: A History*, St Martin's Press, 1999

Boyne, Walter J., *Beyond the Wild Blue: A History of the United States Airforce, 1947–1997*, St Martin's Press, 1998

Braddon, Mary Elizabeth, *Lady Audley's Secret*, 1862, Penguin Classics, 1998

Brody, Aaron, 'Retort Pouches & Trays: A Growing Market', *Food Technology*, Volume 60, Number 4, April 2006

Buffalo Niagara Convention and Visitors Bureau, *Walk Buffalo: A Self-guided Walking Tour of Historic Downtown Buffalo*, 2002

Buford, Bill, *Heat: An Amateur's Adventures as Kitchen Slave, Line Cook, Pasta-Maker, and Apprentice to a Dante-Quoting Butcher in Tuscany*, Knopf, 2006

Bulag, Uradyn E., *Nationalism and Hybridity in Mongolia*, Clarendon Press, 1998

Campbell, George Frederick, *China Tea Clippers*, Adlard Coles, 1974

Clover, Charles, *The End of the Line: How Overfishing Is Changing the World and What We Eat*, Ebury Press, 2004

Cudahy, Brian J., *Box Boats: The Story of Container Ships*, Fordham University Press, 2006

Cullather, Nick, *Secret History: The CIA's Classified Account of Its Operations in Guatemala, 1952–1954*, Stanford University Press, 1999

Darrow, George M., *The Strawberry: History, Breeding and Physiology*, Holt, Rinehart and Winston, 1966

Dart, Joseph, *The Grain Elevators of Buffalo*, paper read before the Buffalo Historical Society, March 13, 1865

Defense Supply Center Philadelphia, Directorate of Subsistence, *Operational Rations Business Unit, Customer Ordering Handbook & Update*, March 2006

Department for Environment, Food and Rural Affairs (Defra), *The Validity of Food Miles as an Indicator of Sustainable Development*, July 2005

Diamond, Jared, *Guns, Germs and Steel: The Fates of Human Societies*, Norton, 1999

Dicke, Thomas S., *Franchising in America: The Development of a Business Method, 1840–1980*, University of North Carolina Press, 1992

Donovan, Arthur, and Bonney, Joseph, *The Box That Changed the World: Fifty Years of Container Shipping – An Illustrated History*, Commonwealth Business Media, 2006

Dore, Giovanna and Nagpal, Tanvi, 'Urban transition in Mongolia: Pursuing Sustainability in a Unique Environment', *Environment*, 1 July 2006

Dulles, Foster Rhea, *The Old China Trade*, Houghton Mifflin, 1930

Earls, Alan R., *US Army Natick Laboratories: The Science Behind the Soldier*, Arcadia Publishing, 2005

Euraque, Daraio A., *Reinterpreting the Banana Republic: Region and State in Honduras, 1870–1972*, University of North Carolina Press, 1996

Evans, John C., *Tea in China: The History of China's National Drink*, Greenwood Press, 1992

Faas, Patrick, *Around the Roman Table*, Macmillan, 2003

Falconer, Bruce, 'US Military Logistics: The Biggest Challenge for the US Armed Forces Today May Be Not the Act of Fighting War but the Feeding, Housing, Arming, Transporting, and Fueling of Their Far-Flung Units',

The Atlantic Monthly, Volume 291, Issue 4, May 2003

Fan, Fa-ti, *British Naturalists in Qing China: Science, Empire and Cultural Encounter*, Harvard University Press, 2004

Fear, A.T., *Rome and Baetica: Urbanization in Southern Spain c. 50 BC–AD 150*, Oxford University Press, 1996

Feige, Chris and Miron, Jeffrey, *The Opium Wars, Opium Legalization, And Opium Consumption in China*, Harvard Institute of Economic Research, Discussion Paper Number 2072, May 2005

Fergusson, Niall, *Empire: How Britain Made the Modern World*, Allen Lane, 2003

Fernandez-Armesto, Felipe, *Food: A History*, Pan Books, 2002

Findling, John E., *Close Neighbors, Distant Friends: United States–Central American Relations*, Greenwood Press, 1987

Flandrin, Jean Louis; Montanari, Massimo and Sonnenfeld, Albert, *Food: A Culinary History from Antiquity to the Present*, Columbia University Press, 1999

Fox, Dennis, *Totally Bananas: The Funny Fruit in American History and Culture*, Xlibris Corp, November 2002

Fratkin, Elliot and Mearns, Robin, 'Sustainability and Pastoral Livelihoods: Lessons from East African Maasai and Mongolia', *Human Organization*, Volume 62, Issue 2, 2003

Gandhi, Mohandas Karamchand, *An Autobiography: The Story of My Experiments with Truth*, Courier Dover Publications, 1983, and Beacon Press, 1993

Gedney, Larry, 'Do Salmon Navigate by the Earth's Magnetic Field?' Article 691, Alaska Science Forum, Geophysical Institute, University of Alaska Fairbanks, 23 November 1984

Gere, Edwin A., *The Unheralded: Men and Women of the Berlin Blockade and Airlift*, Trafford Publishing, 2002

Giangreco, D.M. and Griffin, Robert E., *Airbridge to Berlin: The Berlin Crisis of 1948, its Origins and Aftermath*, Presidio Press, 1988

Gilbert, Bill, *Air Power: Heroes and Heroism in American Flight Missions, 1916 to Today*, Kensington Publishing, 2003

Gleijeses, Piero, *Shattered Hope: The Guatemalan Revolution and the United States, 1944–1954*, Princeton University Press, 1992

Goldman, Mark, *High Hopes: The Rise and Decline of Buffalo*, State University of New York Press, 1983

Greene, Kevin, *The Archaeology of the Roman Economy*, University of California Press, 1990

Griffiths, Percival, *The British Impact on India*, MacDonald, 1952

Griswold, Daniel, *Grain Drain: The Hidden Cost of US Rice Subsidies*, Cato Institute, Center for Trade Policy Studies, November, 2006

Haley, Evan W., *Baetica Felix: People and Prosperity in Southern Spain from Caesar to Septimius Severus*, University of Texas Press, 2003

Hasler, Arthur D. and Wisby, Warren J., 'Discrimination of Stream Odors by Fishes and Its Relation to Parent Stream Behavior', *The American Naturalist*, Volume 85, Number 823, 1951

Hendrickson, John, 'Energy Use in the US Food System: A Summary of Existing Research and Analysis', *Sustainable Farming*, Volume 7, Number 4, Fall 1997

Hines, Colin, *Localization: A Global Manifesto*, Earthscan Publications, 2000

Hines, Colin and Lang, Tim, 'The New Protectionism', *The Nation*, 15 July 1996

Hoffman, Katherine, *Explorations: The Visual Arts Since 1945*, HarperCollins, 1991

Howley, Frank, *Berlin Command*, Putnam, New York, 1950

Hudson, Pat and Hunter, Lynette (editors), 'The Autobiography of William Hart, Cooper, 1776–1857: A Respectable Artisan in the Industrial Revolution', *London Journal*, Volume 7, Number 1, 1981 and Volume 8, Number 1, 1982

Human Rights Watch, *Civilian Devastation, Abuses by All Parties in the War in Southern Sudan*, 1993

Huxley, Elspeth, *The Flame Trees of Thika*, 1959, Penguin Classics, 2000

Immerman, Richard H., *The CIA in Guatemala: The Foreign Policy of Intervention*, University of Texas Press, 1982

Indra, Rinchingiin, 'Mongolian Dairy Products', *Mongolia Today: Science, Culture, Environment and Development*, Badarch, Dendev, Zilinskas, Raymond A., and Balint, Peter J. (editors), Routledge, 2003

Jefferson Loane, Helen, *Industry and Commerce of the City of Rome (50 BC–200 AD)*, The Johns Hopkins Press, 1938

Johnson, Hugh, *Hugh Johnson's Story of Wine*, Octopus Publishing Group, 1989

Jonas, Susanne, *The Battle for Guatemala: Rebels, Death Squads, and US Power*, Westview Press, 1991

Jones, Andy; Hird, Vicki; Longfield, Jeanette and Shub, Merav, *Eating Oil: Food in a Changing Climate*, Sustain, the alliance for better food and farming, and Elm Farm Research Centre, December 2001

Jones, Peter, *Flight Catering*, Elsevier, 2004

Kahn, E.J., Jr, 'A Reporter in Germany', *New Yorker*, 14 May 1949

Karp, David and Lincoln, Murray D., *Vice President in Charge of Revolution*, McGraw-Hill, New York, 1960

Kessler, James H.; Kidd, J.S.; Kidd, Renée A. and Morin, Katherine A., *Distinguished African American Scientists of the 20th Century*, Oryx Press, 1996

Kipling, Rudyard, *American Notes*, R.F. Fenno & Company, Boston, 1899, Kessinger Publishing, 2004

Koehn, Nancy F., *The Power of Commerce: Economy and Governance in the First British Empire*, Cornell University Press, 1994

Kowsy, Francis R., 'Buffalo's Grain Industry and Elevators', in *A History of Buffalo's Grain Elevators*, excerpt from the 2002 Buffalo Grain Elevator Multiple Property Submission to the US National Register of Historic Places

Kulikoff, Allan, *From British Peasants to Colonial American Farmers*, University of North Carolina Press, 2000

Kurlansky, Mark, *Cod: A Biography of the Fish That Changed the World*, Vintage, 1999; *Salt: A World History*, Penguin Books, 2003

Lanciani, Rodolfo, *The Ruins and Excavations of Ancient Rome*, Houghton Mifflin, Boston, 1897

Landes, David S., *The Wealth and Poverty of Nations: Why Some Are So Rich and Some Are So Poor*, Abacus, 1998

Langley, Lester D. and Schoonover, Thomas David, *The Banana Men: American Mercenaries and Entrepreneurs in Central America, 1880–1930*, University Press of Kentucky, 1996

Latham, Ronald Edward (translator), *The Travels of Marco Polo*, Penguin Classics, 1958

Lattimore, Owen, *Nomads and Commissars: Mongolia Revisited*, Oxford University Press, 1962

Lawrence, Felicity, *Not On The Label*, Penguin Books, 2004

Legrain, Philippe, *Open World: The Truth About Globalisation*, Abacus, 2002

Lem, Audun, *China, the WTO and World Fish Trade*, Globefish, Food and Agriculture Organisation

LeRiche, Matthew, *Unintended Alliance: The Co-option of Humanitarian Aid in Conflicts*, Department of War Studies, King's College, University of London, 2004

Levinson, Marc, *The Box: How the Shipping Container Made the World Smaller and the World Economy Bigger*, Princeton University Press, 2006

Lubbock, Alfred Basil, *The China Clippers*, Brown Son & Ferguson, 1946

Luvaas, Jay (editor and translator), *Napoleon on the Art of War*, Touchstone, 1999

Lyding Will, Elizabeth, 'Production, Distribution, and Disposal of Roman Amphoras', *Ceramic Production and Distribution: An Integrated Approach*, Bey, George (editor), Westview Press, 1992; *The Roman Amphora: Learning from Storage Jars*, Archaeology Odyssey, Jan/Feb 2000

Lynn, John (editor), *Feeding Mars: Logistics in Western Warfare from the Middle Ages to the Present*, Westview Press, 1993

Macfarlane, Alan and Macfarlane, Iris, *Green Gold: The Empire of Tea*, Ebury Press, 2004

MacGregor, James and Vorley, Bill, *Fair Miles? The Concept of 'Food Miles' Through a Sustainable Development Lens*, International Institute for Environment and Development, 2006

Mattingly, David, 'First Fruit? The Olive in the Roman World', *Human Landscapes in Classical Antiquity: Environment and Culture*, Salmon, John and Shipley, Graham (editors), Routledge, 1996

May, Stacy and Plaza, Galo, *The United Fruit Company in Latin America*, National Planning Association, 1958

MacGregor, David R., *The Tea Clippers: An Account of the China Tea Trade and of Some of the British Sailing Ships Engaged in it from 1849 to 1869*, Percival Marshall & Co, 1952

McCulloch, Neil and Ota, Masako, *Export Horticulture and Poverty in Kenya*, Institute of Development Studies, University of Sussex, December 2002

McGovern, Patrick E.; Fleming, Stuart J. and Katz, Solomon H. (editors), *The Origins and Ancient History of Wine*, Routledge, 2000

Menor, Larry and Ramasastry, Chandrasekhar, 'Dabbawallahs of Mumbai', *Harvard Business Review*, 26 April 2004

Menzel, Peter and D'Aluisio, Faith, *Hungry Planet: What The World Eats*, Material World Books and Ten Speed Press, 2005

Mili, Samir and Rodríguez Zúñiga, Manuel, 'Exploring Future Developments International Olive Oil Trade and Marketing: A Spanish Perspective', *Agribusiness: An International Journal*, 2001

Mili, Samir and Mahlau, Mario, *Characterization of European Olive Oil*

Production and Markets, European Commission, 2005

Miller, Roger G., *To Save A City: The Berlin Airlift 1948–1949*, US Air Force History and Museums Program, 1998

Millstone, Erik and Lang, Tim, *The Atlas of Food: Who Eats What, Where and Why*, Earthscan Publications, 2002

Morrow Wilson, Charles, *Challenge and Opportunity: Central America*, H. Holt and Company, New York, 1941

Mousseau, Frederic, *Food Aid or Food Sovereignty? Ending World Hunger In Our Time*, The Oakland Institute, 2005

Moxham, Roy, *Tea: Addiction, Exploitation and Empire*, Constable and Robinson, 2003

Murphy, Sophia and McAfee, Kathy, *US Food Aid: Time to Get it Right*, Institute for Agriculture and Trade Policy, Minneapolis, July 2005

Murphy, Sophia, 'The Global Food Basket', *Forum for Applied Research and Public Policy*, University of Tennessee, Volume 16, Issue 2, 2001

Nabhan, Gary Paul, *Coming Home To Eat: Pleasures and Politics of Local Foods*, Norton, 2002

Nestle, Marion, *Food Politics: How the Food Industry Influences Nutrition and Health*, University of California Press, 2003

New York State Writers' Programme, *New York: A Guide to the Empire State*, Oxford University Press, 1940

Paxton, Angela, *The Food Miles Report: The Dangers of Long Distance Food Transport*, SAFE Alliance (now Sustain, the alliance for better food and farming), 1994

Peabody, Erin, 'Fresh-Cut Fruit Moves Into the Fast Lane', *Agricultural Research*, August, 2005

Perdue, Lewis, *Perfect Killer*, Forge, 2005

Pilcher, Jeffrey M., *Food In World History*, Routledge, 2006

Pirog, Rich and Benjamin, Andrew, Leopold Center for Sustainable Agriculture, *Grape Expectations: A Food System Perspective on Redeveloping the Iowa Grape Industry*, April 2000; *Checking the Food Odometer: Comparing Food Miles for Local Versus Conventional Produce Sales to Iowa Institutions*, July 2003; *Calculating Food Miles for a Multiple Ingredient Food Product*, March 2005

Pirog, Rich; Van Pelt, Timothy; Enshayan, Kamyar and Cook, Ellen, *Food, Fuel, and Freeways: An Iowa Perspective on How Far Food Travels, Fuel Usage, and Greenhouse Gas Emissions*, Leopold Center for Sustainable Agriculture, June 2001

Pollan, Michael, *The Omnivore's Dilemma: The Search for the Perfect Meal in a Fast-Food World*, Bloomsbury Publishing, 2006

Powell, Stewart, 'The Berlin Airlift', *Air Force* magazine online, Journal of the Air Force Association, Volume 81, Number 6, June 1998

Robinson, Jancis, *The Oxford Companion to Wine*, Oxford University Press, 2006

Rodríguez, José Remesal, *The Contribution of Mount Testaccio Towards the Understanding of the Roman Economy in the Empire*, conference presentation at the opening of Monte Testaccio exhibition by Berni, P.; Cubero, M. and Aguilera, Antonio, University of Rome 'La Sapienza', 15 January 1997

Rosenblum, Mort, *Olives: The Life and Lore of a Noble Fruit*, North Point Press, 1996

Saad, Rana, *William of Rubruck's Account of the Mongols*, Lulu Press, 2004

Samuel, Wolfgang W. E., *I Always Wanted to Fly: America's Cold War Airmen*, University Press of Mississippi, 2001; *The War of Our Childhood: Memories of World War II*, University Press of Mississippi, 2002

Sands, William Franklin and Lalley, Joseph M., *Our Jungle Diplomacy*, University of North Carolina Press, 1944

Saunders, Caroline; Barber, Andrew and Taylor, Greg, *Food Miles: Comparative Energy/Emissions Performance of New Zealand's Agriculture Industry*, Research Report Number 285, Lincoln University, July 2006

Scharlin, Patricia and Taylor, Gary, *Smart Alliance: How a Global Corporation and Environmental Activists Transformed a Tarnished Brand*, Yale University Press, 2004

Schlesinger, Stephen and Kinzer, Stephen, *Bitter Fruit: The Story of the American Coup in Guatemala*, Harvard University Press, 2005

Schlosser, Eric, *Fast Food Nation: What the All-American Meal Is Doing to the World*, Penguin Books, 2002

Schneekloth, Lynda (editor), *Reconsidering Concrete Atlantis: Buffalo Grain Elevators*, The Urban Design Project, School of Architecture and Planning, University at Buffalo, State University of New York, 2006

Scidmore, Eliza Ruhamah, *Winter India*, The Century Co, 1904

Scott Jenkins, Virginia, *Bananas: An American History*, Smithsonian Institution Press, 2000

Shaw, D. John, *The UN World Food Programme and the Development of Food Aid* Palgrave, 2001

Shephard, Sue, *Pickled, Potted and Canned: The Story of Food Preserving*, Headline Book Publishing, 2000

Shipping Wonders of the World: Romance of the Seven Seas in Story and Pictures, the Amalgamated Press, London, 1936

Singer, Peter and Mason, Jim, *Eating: What We Eat and Why It Matters*, Arrow, 2006

Sneath, David, *Notions of Rights over Land and the History of Mongolian Pastoralism*, presentation at the Eighth Conference of the International Association for the Study of Common Property, Bloomington, Indiana, May 2000

Stein, Nicholas, 'Yes, We Have No Profits: The Rise and Fall of Chiquita Banana: How a Great American Brand Lost its Way', *Fortune*, 26 November 2001

Streeter, Stephen M., 'Interpreting the 1954 US Intervention in Guatemala: Realist, Revisionist, and Postrevisionist Perspectives', *The History Teacher*, Volume 34, Number 1, November 2000

Sundloff Schulz, Deborah and Schulz, Donald E., *The United States, Honduras and the Crisis in Central America*, Westview Press, 1994

Sustain, the alliance for better food and farming, *Food Miles – Still on the Road to Ruin? An Assessment of the Debate over the Unnecessary Transport of Food, Five Years on from the Food Miles Report*, October 1999

Sutie, J.M., *Country Pasture/Forage Resource Profiles: Mongolia*, Food and Agriculture Organisation, 2000

Swedish International Development Co-operation Agency, *An Evaluation of the Programme Export Promotion of Organic Products from Africa, Phase II*, October 2004

Tannahill, Reay, *Food in History*, Eyre Methuen, 1973
The Indian Mirror or *Illustrations of Bible Truth Drawn from Life in India*, Thomas Nelson and Sons, London, 1878
Toussaint-Samat, Maguelonne, *History of Food*, Blackwell Publishing, 1992
Tunner, William, *Over the Hump*, Duell, Sloan and Pearce, New York, 1964
Turner, Jack, *Spices: The History of a Temptation*, Harper Perennial, 2005
Tusa, Ann and John, *The Berlin Airlift*, Atheneum, 1988
Twede, Diana, 'The Packaging Technology and Science of Ancient Transport Amphoras', *Packaging, Technology and Science*, Michigan State University, 2002; 'Commercial Amphoras: The Earliest Consumer Packages?' *Journal of Macromarketing*, Volume 22, Number 1, June 2002; *Basket, Barrel and Box: the Secret Role of Crafted Wooden Containers in the History of Marketing*, Michigan State and University School of Packaging, 2003
UN-Habitat, United Nations Human Settlements Programme, *State of the World's Cities 2006/7*, 2006
United Nations Food and Agriculture Organisation with the Livestock, Environment and Development Initiative, *Livestock's Long Shadow: Environmental Issues and Options*, 2006
Unwin, Tim, *Wine and the Vine: An Historical Geography of Viticulture and the Wine Trade*, Routledge, 1996
US Agency for International Development, *Celebrating Food for Peace, 1954–2004: Bringing Hope to the Hungry*, USAID
Van Crevald, Martin L., *Supplying War: Logistics from Wallenstein to Patton*, Cambridge University Press, 1979
Visser, Margaret, *Much Depends on Dinner: The Extraordinary History and Mythology, Allure and Obsessions, Perils and Taboos, of an Ordinary Meal*, Penguin Books, 1989
Wellfleet Non-resident Taxpayer's Association Newsletter, 'Wellfleet's most famous non-resident', Summer 2005, Volume IV, Number 1
Werner, Emmy, *Through the Eyes of Innocents: Children Witness World War II*, Westview Press, 2000
Wilson, C. Anne (editor), *Waste Not, Want Not: Food Preservation from Early Times to the Present Day*, Edinburgh University Press, 1991

a note on sources

In the process of researching this book, I read numerous articles in newspapers and magazines. Some are listed in this bibliography but certain publications proved invaluable. These include *The Economist*, the *Financial Times*, the *Guardian* and the *New York Times*. Useful websites include carbontrust.co.uk (Carbon Trust), bbc.co.uk (BBC News and *The Food Programme*, Radio 4), epa.gov (US Environmental Protection Agency), fao.org (United Nations Food and Agriculture Organisation), food.gov.uk (Food Standards Agency), foodnavigator.com (a Decision News Media newsletter), usaid.gov (United States Agency for International Development), wflo.org (World Food Logistics Organization), wfp.org (World Food Programme) and wri.org (World Resources Institute). For more details go to moveablefeasts.org

index

agribusiness 58, 261, 296
agriculture
  collective farming 169–70
  early 312
  subsidies 56, 58, 311
  surpluses 58, 295, 312
air transport
  belly-hold freight 247–8, 249–51, 252
  Berlin Airlift 83–105
  food aid programmes 286–91, 293–4,
    299–300, 305–6
airag 162, 163, 172, 174
airline meals 243–7
Alesund 37
American Civil War 68, 78, 112, 266
American eating habits 120
amphorae 8, 9, 11–12, 14, 15–16, 17–18,
  20, 30, 196, 281
Andalusia 8–9, 22–3
Antonov aircraft 286–8
anti-globalisation movement 53, 58–9
Appert, Nicholas 73–5, 76, 77, 78
Arbenz Guzmàn, Jacobo 135, 136, 137,
  154, 156, 157, 158, 159
Arbroath smokies 28
Arévalo Bermejo, Juan José 153–4
Armas, Carlos Castillo 157, 158
art and architecture 265, 277–84
artificial flavours 210
astronauts' diets 239–40
Aungier, Gerald 114
avocados 50

Back, Charles 29–30
Baetica 8, 9, 12, 16, 17, 18, 19, 30
Baker, Captain Lorenzo Dow 135, 136,
  138–9, 158
balsamic vinegar 4, 201
banana republics 4, 137, 152
bananas 4, 39, 54, 134, 136–45, 146–51,
  157, 159
Banderas, Antonio 24
Bangladesh 54–5
Banham, Reyner 275–6, 277–8, 280
barcodes 125–6, 129
barrels, wooden 4, 192–3, 194–201,
  205–9, 281
  construction 195–6, 200–1
  oak adjuncts 4, 207–8
Barrett, Christopher 295, 296
Beames, John 110

beer 37, 205–7
Beeton, Isabella 224
Berlin Airlift 83–105, 160, 288, 295
Berlin Wall 104
Bernays, Edward 155
Bertolli 21–2
Bertrand, Francisco 153
Bettys & Taylors of Harrogate 228, 231
biofuels 189, 258
Bioregenerative Life Support Systems
  project 240
blood, drinking 171–2
Bocas del Toro 143–4
Böge, Stefanie 177, 180, 308
Bonilla, Manuel 152, 153
Boscawen, Hon. Evelyn 234, 236
Boston Fruit Company 136, 139, 141, 150
Bové, José 53–4
Boyle, Robert 75
Braddon, Mary Elizabeth 222
Branson, Richard 131
break-bulk shipping 39–40
Brisbane 45–6
Brussels sprouts 27
Bruzza, Father Luigi 14
Buffalo, New York 5, 263–5, 266–80, 315
butter 181

cans 73, 77, 78–9, 80
car companies 52
carbon dioxide emissions 185–6, 187,
  257–8, 260, 309, 310, 313
carbon footprint 187
Care packages 91–3, 102
carob gum 180
caste system 115–16
Celts 196
Central Intelligence Agency (CIA) 156–7,
  158, 159
champagne 28, 29
Charlemagne 70
Charles, Prince of Wales 109, 228
Cheddar cheese 27
cheese 26, 27–8, 29
Chen, Keith 227–9, 232–3
chewing gum 82, 93–4
child labour 54–5
China 25, 36, 37–8, 51–2, 214, 215, 220,
  223, 225–6, 227–33
Chiquita 54, 55, 136, 158
Clay, General Lucius 86, 87–8

clipper races 213, 215–17, 218–20, 221–2, 227, 282
clothing and footwear companies 52, 54, 55
coffee 89, 185
Colbert, Gregory 283
Cold War 156, 295
Coleridge, Samuel Taylor 225
Common Agricultural Policy 58
communism 85, 86, 155–6, 157, 158, 159, 160, 169–70, 214, 295
Compressed Meal (CM) 64
container shipping 3, 31, 37, 38–44, 46–51, 55–6, 58, 257, 282–3
    reefer units 47–9, 50–1, 137
    security 40–1, 42
contrails 257
cooking equipment 71–2
coopers 192, 195, 196, 197, 198–9, 201
corn on the cob 308–9
Cornelius, Cole 207
Cornish pasties 111
Corridan, Father John 45
crab 27, 52
cruise business 148–9
curry 106
Cutty Sark 213, 219

dabbawallahs 108, 110, 113, 117–19, 130, 131–2, 133
dairy foods 164, 172, 173–6, 187, 232
Daoudi, Amer 300–1, 302
Darsch, Gerry 61, 69, 70
Dart, Joseph 5, 265–6, 269–70, 271, 273, 312–3
Dávila, Miguel 152
Dawson, Montague 282
deforestation 185, 189, 260
dehydrated foods 89
demographic shift 187–8, 312
Dickens, Charles 44–5
dockworkers 44–5, 46
dollar diplomacy 137, 145, 152
Domino's 121–3
Donkin, Brian 77, 78
Dressel, Heinrich 14
Dulles, Allen 159
Dulles, Foster Rhea 217
Dulles, John Foster 159
Dunbar, Robert 265–6, 271
Durand, Peter 77, 78

East India Company 222, 224–5, 227
eggs 194
Emma Maersk 41–2, 187, 310, 313
energy usage 185, 187
environmental degradation 185, 260–1
Erie Canal 268–9, 279
estufagem 204

ethanol 189–90, 258, 314–15
European Union 28, 29, 299
Evans, Oliver 271
export subsidies 23

farmers' markets 182–3
'farmhouse foods' 25
FedEx 124–30, 301
Fertile Crescent 312
feta cheese 28, 29
fictional locations 25
First Strike Ration (FSR) 64, 65
fish oils 179–80
fish paste, Roman 3
flour 89, 93, 100, 262
foie gras 25
food aid programmes 286–306
    commercial partnerships 301–2
    GM food aid 298–9
    inefficiencies 295, 296–7, 299–300
    logistical obstacles 302
    misappropriation 298
    tonnage 303
food dependency culture 299, 303
food miles 164–5, 180, 181–2, 258, 308, 309
food preservation 74–80
Fortune, Robert 227, 235
forward osmosis 64
Frankfurt airport 249–50
free trade 57
French beans 2, 27
'fresh cut' food 253–4
fructose 180
fruit bottling 75
fruits, preserved 30, 32
fuel efficiency 314–15

Gabriel, Richard 68
Gama, Vasco da 1, 4, 25
genetically modified foods 298–9
Genghis Khan 171, 173
geographical indication system 28–9
ger 163, 167, 171, 172
Ghana 258–9
Giuliani, Carlo 59
Glasse, Hannah 75
global branding 55
global positioning systems 51
global warming 181, 309
globalisation 56
goat 163
Goldfinger, Erno 280
Gordon, George James 227
gourmet meal deliveries 123–5
grain 19, 30, 70, 71, 266, 268, 269–71, 302
grain elevators 5, 263, 264–5, 271–80, 283, 284, 315

grapes 2, 183, 249
Great White Fleet 148, 154
Greek terracotta 280–1
greenhouse gases 185, 186, 257, 309
Grimod de la Reynière, Alexandre
    Balthazar 74
Gropius, Walter 265, 278, 279
Guadalquivir river 18
Guatemala 135, 145, 150, 151, 153–8,
    159–60

Hall, John 77, 78
Halvorsen, Gail 93–4
Hamburg 45
Hart, William 198, 199, 200
Hasler, Arthur 35
Hazlitt, William 225
health and safety regulations 38,
    209–10
Heathrow airport 250
Hines, Colin 183
Honduras 143, 151, 152–3
Hooah! bar 80–1
Howell, Bill 124
Howley, Frank 87–8
hub-and-spoke system 130
Humphrey, Hubert 295
hunter-gathering 312
Huxley, Elspeth 171–2

ice cream 51
import taxes 56, 58, 152, 153
Incas 310–11
India 25, 57, 108–10, 112–20, 132–4, 198,
    223–4, 227
India Pale Ale 206–7
Indra, Rinchingiin 177
integrated transport systems 43–4
intermodalism 38, 41, 43
International Style 284
internet 54
irradiated food 239–40

Jaén 23
Jamaica 138, 139, 148–9
jams and marmalades 78
Japan 4
Jardine Matheson 225
Jones, Frederick 49–50
Jones, Jonathan 235–7

Keay, John 213, 216
Keith, Minor 141–2, 150
Kenya 2, 259, 261, 311
Khaki Weed 71
Kipling, Rudyard 272
kiwi fruit 50, 140, 249
Korea 56, 57
Kublai Khan 162

labour costs 41, 51–2, 56
lactic acid 175–6
Lamikanra, Olusola 254
Lanciani, Rodolfo 13
land acquisition 151, 154–5
Lang, Tim 183
Le Corbusier 277, 278, 280, 284
Lee Kyung-hae 53, 56
'lettuce crisis' 253
lobster 78
locally produced food 180, 182–3, 184,
    185, 188, 258, 299, 309, 310, 313
Lokichoggio 291–3, 303–5
London docklands 46, 198–200, 283
Lucas, Caroline 183
lycopene 79

McDonald's 116
Mclean, Malcolm 39, 41, 48
Madeira wines 203–5, 207
maize 285, 286, 291, 293, 294–5, 297–8
malnutrition 290
mangoes 140, 307
Masai people 171–2
Mason, Jim 184, 185, 186, 258
Maxwell, Daniel 295
Mayhew, Henry 44
Meal, Alternative Regionally Customized
    (MARC) 69–70
Meals Ready to Eat (MREs) 63, 64–5,
    68–9, 73
meat shipments 145, 146, 181
Medge, Raghunath 109–10, 118, 119,
    132
Melton Mowbray pork pies 28, 30
Mendelsohn, Erich 265, 277
metabolic dominance 67
metacities 188
methane 186
microbial cultures 175–6
microprocessors 50, 51
milk 173–4
milking process 175
mitochondrial function 67, 68
modernist architecture 265, 277–8,
    279–80, 284
modified atmosphere packaging (MAP)
    254
molasses 198
Mombasa 297–8
Monaghan, James and Tom 121–2
Mongolia 162–76, 186, 188, 190
Monte Testaccio 9–14, 15–16, 17
Mozambique 311
mozzarella 26
Mumbai 108–10, 112–14, 117–20, 132–4
Mundy, Gerry 252, 253
Musa, Sextus Fadius Secundus 17
mutton 162, 163, 181

Nabhan, Gary Paul 182
Napoleon Bonaparte 76–7, 78
NASA 239–40
Natick Soldier Systems Centre 61–2, 65,
    67, 79
Needham, John 75
Newcastle brown ale 28
nitrous oxide emissions 186
Nomadic Museum 283
nomadism 165, 167–9, 170, 171
North Atlantic Treaty Organisation
    (NATO) 104
Numero, Joseph 49, 50
nuts 30, 32

olive oil 5–6, 7, 15, 18–19, 20–4, 26, 30,
    31, 308
  blended 22, 23
  marketing 21–2, 23–4
olive oil tankers 31, 209
olive trees 23
Olmstead, Frederick Law 267
omega-3 179–80
On the Waterfront 45
Operation Lifeline Sudan 286, 287–91
Operation Little Vittles 94
Operation Plainfare see Berlin Airlift
Operation Vittles see Berlin Airlift
opium 223–6
Opium Wars 226
oranges 25, 26
organic food 258, 259
Orwell, George 214
Ostia 19–20, 31–2
Oswald, Ron 158
over-grazing 171
oysters 78

Papin, Denis 75
Parmigiano Reggiano 28
passion fruit 140
Pasteur, Louis 74
pasteurization 74
pastoralism 165, 166, 168, 171
Peak Soldier Performance Programme
    67
pepper 1, 2, 4, 31
Perdue, Lewis 67, 68
Piero della Francesca 281
pigs 171
pineapples 255, 258
pizza delivery service 121–3
Pliny the Elder 18
Polo, Marco 166–7, 173
polytunnels 256, 257
Portland, Oregon 188
Pratt, Barbara 48–9, 50
Preston, Andrew 139, 141, 142, 150
product names, protecting 27–30

Prosciutto di Parma 28
protected designation of origin (PDO) 28
protected geographical indication (PGI)
    28–9
protectionism 56, 57–8
Pumpelly, Raphael 202–3

Qing dynasty 214
Qingdao 37, 52, 56

railways 150, 152
Rainforest Alliance 55, 153, 158
rations
  food aid programmes 290, 294
  portable (military) 62–76, 76
  wartime 88–90, 102
reefer units 47–9, 50–1, 137
refrigerated containers see reefer units
refrigerated steamships 4, 136, 145–9,
    151, 158
religious dietary restrictions 114–15
Remesal Rodríguez, José 8, 9, 10, 12
retort packaging 65–6, 72–3, 79–80
rice 56, 57, 58, 185, 198
ripening process 51
robotics 46
Rogers, Bob 207
Romans 2–3, 9, 15, 18, 22–3, 30, 196–7
Rome 8–17
rooftop gardens 314
routing technology 314
Rowe, Ollie 182, 313

Saint Laurent, Yves 29
Saint Wandrille monastery 70
Salisbury, Val 255, 256
salmon 33, 308
  migration 34–6
  processing 36–8
Salmon, Robert 282
salt 95, 291
sandwiches 111
satellite technology 51
scanning and tracking devices 40–1
Schuffert, John 99
Schuyler, General Philip 71
Scidmore, Eliza Ruhamah 110
Sepoy Rebellion (1857) 115
Shanghai 226, 229–32
Sharp, Dougal 205, 206
sheep 162, 163, 168, 181
shelf life 51, 72, 252
ships
  refrigerated steamships 4, 136, 145–9,
    151, 158
  Roman 19
  unloading 42, 221, 270–1
  see also clipper races; container
    shipping

shrinkage rates 146
Silk Road 2
Singer, Peter 184, 185, 186, 258
Six Sigma quality control 107–8, 131
slave trade 198
Slow Food movement 184
Smith, Adam 183, 235
Smith, Fred 129–30
Smith, Rick 315
smoked salmon 25
sorting machinery, automatic 128–9
Southers, Doug 100, 101–2
soya 185, 189
Spain 6, 8–9, 22–4, 253
Spallanzani, Lazzaro 75
Sparrow, Andy 244, 245
spice trade 1–2, 4, 25, 30
Stalin, Joseph 84, 85, 86, 89, 103, 156
Steele, Robert 219
sterilisation 72–3, 74, 77, 78
Stewart, Heather 304
strawberries 177, 178, 238, 240–3, 247–8,
    249, 250–2, 255–7, 308
strawberry festival 242
Sudan 286, 287–94, 298, 299, 300, 303–6,
    311–12
sugar 27, 189, 198, 204, 291
supply chain management 131, 132–3
Sustain 180

Tallien, Madame 241–2
Tanzania 259
Taut, Bruno 277, 279
tea 211, 212–15, 216, 220–3, 224, 225,
    226–37, 308
    caravan tea 201–3
    Darjeeling 28, 235
    English tea plantation 234–6, 237
    milk tea 232
    rose petal tea 231
teahouses 230
Tempelhof 93, 94, 96, 97–8, 100–1
temperance movement 223
Tetra Pak 80
Thermo King 50
Thermopylae 212, 213, 216
Thomas, Gerald 120
tiffin 110–11
tiffin delivery service 108–9, 112–13,
    116–20, 130, 131–2, 133
Toltz, Max 275
tomato juice 245
tomatoes 26, 60, 79, 80, 184, 308
trade barriers 29
tradition speciality guaranteed (TSG) 29
transdermal nutrient delivery system
    (TDNDS) 66–7

transhumance *see* pastoralism
transport efficiencies 313–14
Tregothnan 234–6, 237
Trollope, Anthony 265, 272, 276
truffles 25
tuna 36
Tunner, General William 96–9, 103,
    104–5
turkeys 123–4
TV dinners 120–1, 122

Ulaan Baatar 163, 188
ultra-high-pressure processing 72
UN World Food Programme 286–94,
    299, 300–3, 305
United Fruit Company 141, 142, 143,
    144–5, 146, 148–9, 150, 151, 152,
    153, 154–5, 158, 159
USAID 258–9, 294, 296–8

vertical integration 150
Vietnam 57

Wait, Harry 275
Walter, Lionel 209
Walters, Samuel 282
Wardian cases 234–5
Warhol, Andy 282
water, bottled 25–6
watermelons 4
Webber, Mike 205
Welland Canal 278–9
Werner, Harry 49
whisky 42, 205
Wiggin, Bill 256
William of Rubruck 169, 174
Williams, Helen Maria 240–1
Wilson, Charles Morrow 149–50
Wimbledon week 241
wine and winemaking 4, 25, 29–30, 184,
    191, 193–4, 308
    altitude effects 244–5
    bouquet 193
    Madeira wines 203–5, 207
    oak-aged 4, 200–1, 207
Wisby, Warren 35
Wolf, Martin 184
working conditions 54, 55, 56, 153–4,
    158, 259
World Trade Organisation 29, 53

yaks 168, 174
yogurt 161, 164–5, 173, 176–80, 308

Zambia 260, 298, 299
Zemurray, Sam 142, 152, 153